OPTICAL BIT ERROR RATE

IEEE Press
445 Hoes Lane
Piscataway, NJ 08854

IEEE Press Editorial Board
Stamatios V. Kartalopoulos, *Editor in Chief*

M. Akay	M. E. El-Hawary	F. M. B. Pereira
J. B. Anderson	R. Leonardi	C. Singh
R. J. Baker	M. Montrose	S. Tewksbury
J. E. Brewer	M. S. Newman	G. Zobrist

Kenneth Moore, *Director of IEEE Book and Information Services (BIS)*
Catherine Faduska, *Senior Acquisitions Editor*
Anthony VenGraitis, *Project Editor*

OPTICAL BIT ERROR RATE
An Estimation Methodology

STAMATIOS V. KARTALOPOULOS

IEEE PRESS

A JOHN WILEY & SONS, INC., PUBLICATION

The content of this book is intended to have illustrative and educational value and it should not be regarded as a complete specification for the design of optical networks. The information presented in this book is adapted from standards and from the author's research activities; however, although a serious effort has been made, the author does not warranty that changes have not been made and typographical errors do not exist. The reader is encouraged to consult the most current standards recommendations and manufacturer's data sheets.

Copyright © 2004 by the Institute of Electrical and Electronics Engineers, Inc. All rights reserved.

Published simultaneously in Canada.

No part of this publication may be reproduced, stored in a retrieval system or transmitted in any form or by any means, electronic, mechanical, photocopying, recording, scanning or otherwise, except as permitted under Section 107 or 108 of the 1976 United States Copyright Act, without either the prior written permission of the Publisher, or authorization through payment of the appropriate per-copy fee to the Copyright Clearance Center, Inc., 222 Rosewood Drive, Danvers, MA 01923, (978) 750-8400, fax (978) 646-8600, or on the web at www.copyright.com. Requests to the Publisher for permission should be addressed to the Permissions Department, John Wiley & Sons, Inc., 111 River Street, Hoboken, NJ 07030, (201) 748-6011, fax (201) 748-6008.

Limit of Liability/Disclaimer of Warranty: While the publisher and author have used their best efforts in preparing this book, they make no representation or warranties with respect to the accuracy or completeness of the contents of this book and specifically disclaim any implied warranties of merchantability or fitness for a particular purpose. No warranty may be created or extended by sales representatives or written sales materials. The advice and strategies contained herein may not be suitable for your situation. You should consult with a professional where appropriate. Neither the publisher nor author shall be liable for any loss of profit or any other commercial damages, including but not limited to special, incidental, consequential, or other damages.

For general information on our other products and services please contact our Customer Care Department within the U.S. at 877-762-2974, outside the U.S. at 317-572-3993 or fax 317-572-4002.

Wiley also publishes its books in a variety of electronic formats. Some content that appears in print, however, may not be available in electronic format.

Library of Congress Cataloging-in-Publication Data is available.

ISBN 0-471-61545-5

Printed in the United States of America.

10 9 8 7 6 5 4 3 2 1

To a world in which errors can be predicted, detected, and corrected to better than 1 in 100,000,000,000,000

Contents

Preface		xi
Acknowledgments		xix
Constants, Conversions and Useful Formulae		xxi
Introduction		1
1	**Principles of Modulation and Digital Transmission**	**7**
1.1	Digital Versus Analog	7
1.2	Spectrum in Optical Communications	9
	1.2.1 ITU-T Nominal Center Frequencies	9
1.3	Linear Response to Square Input Pulses	10
1.4	Principles of Modulation	11
	1.4.1 On-Off Keying and NRZ Versus RZ	13
	1.4.2 A Case of Amplitude Modulation	15
	1.4.3 Phase-Shift Keying	16
	1.4.4 Frequency-Shift Keying	17
	1.4.5 Duo-Binary Modulation	17
	1.4.6 State-of-Polarization Shift Keying	19
	1.4.7 Comparison between OOK and PSK	20
1.5	Modulator Types	20
1.6	Principles of Decoding	21
	References	24
	Standards	25
2	**Optical Propagation**	**27**
2.1	Introduction	27
2.2	The Wave Nature of Light	33
2.3	Classical Interference	33
2.4	Quantum Interference	34
2.5	Light Attributes	35
2.6	Matter	35
	2.6.1 Attributes of Matter	36

2.7	Propagation of Light	37
	2.7.1 Reflection and Refraction—Snell's Law	37
	2.7.2 Phase and Group Velocity	39
	2.7.3 Spectral Broadening	40
	2.7.4 Scattering and Absorption	40
	2.7.5 Microcracks	40
	2.7.6 Mechanical Pressure	41
	2.7.7 Temperature Variation	41
2.8	Diffraction	41
2.9	Polarization	44
	2.9.1 The Stokes Vector	47
	2.9.2 The Poincaré Sphere	49
	2.9.3 Birefringence	50
	2.9.4 Polarization-Dependent Loss	50
	2.9.5 Extinction Ratio	51
	2.9.6 Phase Shift	52
2.10	Paradoxes	52
	2.10.1 Background	52
	2.10.2 Paradoxes	53
2.11	Material Dispersion	53
2.12	Glass Fiber, an Optical Transmission Medium	54
	2.12.1 Fiber Modes	55
	2.12.2 Multimode Fiber	56
	2.12.3 Single-Mode Characteristics	56
	2.12.4 Effective Area	57
	2.12.5 Cut-Off Wavelength	58
	2.12.6 Fiber Attenuation and Power Loss	58
	2.12.7 Limits of Optical Power in Fiber	60
	2.12.8 Fiber Birefringence	61
2.13	Dispersion	62
	2.13.1 Modal Dispersion	62
	2.13.2 Chromatic Dispersion	64
	2.13.4 Dispersion-Shifted Fiber	67
	2.13.5 Dispersion Slope, Dispersion Compensation, and Figure of Merit	68
	2.13.6 Polarization-Mode Dispersion	69
	2.13.7 Polarization-Mode Dispersion Compensation	70
	2.13.8 Polarization-Mode Coupling	70
	2.13.9 Pulse Timing and Dispersion	70
2.14	Fiber Polarization-Dependent Loss	71
2.15	Self-Phase Modulation	71
2.16	Self-Modulation or Modulation Instability	73
2.17	Effect of Pulse Broadening on Bit Error Rate	73
2.18	Four-Wave Mixing	73
2.19	The Decibel Unit	75
	References	78

3 Optical Transmitters and Receivers — 81

- 3.1 Introduction — 81
- 3.2 The Transmitter — 81
  - 3.2.1 Lasers — 83
- 3.3 The Receiver — 84
  - 3.3.1 Detection — 86
  - 3.3.2 Photodetectors — 91
  - 3.3.3 PIN Photodiodes — 93
  - 3.3.4 APD Photodiodes — 94
  - 3.3.5 Photodetector Figure of Merit — 95
- 3.4 Detection Techniques — 96
  - 3.4.1 Demodulation of the Optical Signal — 97
    - 3.4.1.1 OOK RZ or NRZ — 97
    - 3.4.1.2 PSK and FSK — 97
  - 3.4.3 Pulse-Shape — 98
  - 3.4.4 Sampling — 98
  - 3.4.5 Sampling Methods — 99
- References — 99

4 Overview of DWDM Devices and Networks — 101

- 4.1 Introduction — 101
- 4.2 Review of DWDM Components — 102
  - 4.2.1 The Fiber in Optical DWDM Transmission — 102
  - 4.2.2 Optical Multiplexers and Demultiplexers — 104
  - 4.2.3 Optical Filters — 104
  - 4.2.4 Rotators—The Faraday Effect — 107
  - 4.2.5 Optical Power Attenuators — 107
  - 4.2.6 Optical Isolators and Circulators — 108
  - 4.2.7 Gratings — 109
    - 4.2.7.1 Fiber Bragg Gratings — 112
    - 4.2.7.2 Dependence of Gratings on Temperature — 112
  - 4.2.8 Optical Amplifiers — 113
    - 4.2.8.1 Semiconductor Optical Amplifiers — 114
    - 4.2.8.2 Optical Fiber Amplifiers — 115
    - 4.2.8.3 Amplified Spontaneous Light Emission — 117
    - 4.2.8.4 Raman Amplifiers — 117
    - 4.2.8.5 Raman Noise — 119
    - 4.2.8.6 OFA and Raman Amplifiers — 121
  - 4.2.9 Stimulated Brillouin Scattering — 122
  - 4.2.10 Wavelength Converters — 122
    - 4.2.10.1 Cross-Gain Modulation — 123
    - 4.2.10.2 Cross-Phase Modulation — 123
- 4.3 Interaction of Multiple Channels in Fiber — 123
  - 4.3.1 Nonlinear Phenomena — 124

		4.3.2	Four-Wave Mixing	124
		4.3.3	Temporal FWM	126
		4.3.4	Impact of FWM on DWDM Transmission Systems	127
		4.3.5	Summary of Nonlinear Phenomena	127
		4.3.6	Factors Affecting Matter and Light	128
	4.4	Review of DWDM Networks		128
		4.4.1	Next-Generation SONET/SDH	130
		4.4.2	Topologies	131
		4.4.3	The Optical Transport Network	133
		4.4.4	DWDM Network Restoration	134
		4.4.5	Bandwidth Management	136
		4.4.6	Wavelength Management	136
		4.4.7	Optical-Performance Monitoring	137
		References		138
		Standards		144

5 Noise Sources Affecting the Optical Signal — 147

	5.1	Introduction		147
		5.1.1	What is Noise?	147
		5.1.2	Sources of Noise	147
		5.1.3	Oscillator Noise	149
		5.1.4	Statistical Noise	149
	5.2	Transmission Factors		150
		5.2.1	Phase Distortion	150
		5.2.2	Frequency Distortion	151
		5.2.3	Polarization Distortion	152
		5.2.4	Echo	152
		5.2.5	Singing	153
	5.3	Thermal Noise		153
		5.3.1	Thermal Noise at the Receiver	158
	5.4	Shot Noise		159
		5.4.1	Electronic Shot Noise	159
		5.4.2	Shot Noise and the Dual Nature of Photons—The Fano Factor	161
		5.4.3	Shot Noise—Conclusion	162
	5.5	Flicker or 1/f Noise		162
	5.6	Other Noise Sources		164
		5.6.1	Amplifier Noise	164
		5.6.2	Four Wave Mixing	166
		5.6.3	Cross Talk	166
		5.6.4	Polarization-Mode Dispersion	167
		5.6.5	Stokes Noise and Chromatic Jitter	167
		5.6.6	Spectral Broadening	168
		5.6.7	Self-Phase Modulation	168

	5.6.8	Self-Modulation or Modulation Instability	169
	5.6.9	Attenuation	169
5.7	Temporal Parameters		169
5.8	Linear, Nonlinear, and Other Effects		170
5.9	Photodetector Responsivity and Noise Contributors		173
	References		175

6 Timing, Jitter, and Wander   181

6.1	The Primary Reference Source		181
6.2	The Phase-Lock Loop		182
6.3	Bit, Frame and Payload Synchronization		184
	6.3.1	SONET Synchronization—An Example	185
6.4	Synchronization Impairments		188
6.5	Network Hierarchy		189
6.6	Theoretical Foundation of Timing Error		190
	6.6.1	Maximum Time Interval Error	192
	6.6.2	Time Deviation	192
	6.6.3	Maximum TIE	194
	6.6.4	Allan Deviation	194
	6.6.5	Comparison of Timing Models Defined in G.810	195
6.7	Traffic Probability and Signal Quality		195
6.8	Jitter and Wander		196
	6.8.1	Types of Jitter	199
		6.8.1.1 Intersymbol Interference (ISI)	200
		6.8.1.2 Data-Dependent Jitter (DDJ)	200
		6.8.1.3 Pulse-Width Distortion Jitter (PWDJ)	200
		6.8.1.4 Sinusoidal Jitter (SJ)	201
		6.8.1.5 Uncorrelated Bounded Jitter (UBJ)	201
	6.8.2	Signal Affected by Jitter	201
	6.8.3	Sources of Jitter	201
	6.8.4	Examples of Self-Inflicted Optical Noise and Jitter: Stokes Noise and Chromatic Jitter FWM	202
	6.8.5	Examples of WDM Optical Noise and Jitter: Four-Wave Mixing	203
	6.8.6	Jitter Generation, Tolerance, and Transfer	203
	6.8.7	Jitter Filtering Templates	204
	6.8.8	Maximum Jitter Tolerance Templates	205
	6.8.9	Bit Error Ratio Penalty Criterion	206
6.9	Wander		207
	6.9.1	Sources of Wander	207
	6.9.2	Signal Affected by Wander	207
	References		209
	Standards		210

7 Probability Theory of Bit Error Rate — 213

- 7.1 Introduction — 213
- 7.2 Bit Error Ratio and Bit Error Rate — 214
- 7.3 Definitions — 215
- 7.4 OSNR and Spectral Matching — 216
- 7.5 Carrier-to-Noise Ratio — 217
- 7.6 Shannon's Limit — 219
- 7.7 Optical Signal-to-Noise Ratio — 219
- 7.8 Probability and Statistics 101 — 220
  - 7.8.1 Binomial Distribution — 224
  - 7.8.2 Gaussian and Normal Distributions — 225
  - 7.8.3 Poisson Distribution — 225
- 7.9 Bit Error Probability — 226
- 7.10 Bit Error Contributors — 227
  - 7.10.1 Optical Nonlinearities — 227
  - 7.10.2 Polarization — 228
  - 7.10.3 Other Factors — 228
- 7.11 Bit Error Rate — 228
- 7.12 Optical Signal-to-Noise Ratio — 233
- 7.13 Channel Capacity and SNR — 234
- 7.14 The Quality or Q-Factor — 237
- 7.15 Bit Error Monitoring — 239
- 7.16 BER and Data Patterning — 239
- 7.17 BER and Eye Diagram — 240
- 7.18 Eye-Pattern Mask — 243
- References — 245
- Standards — 247

8 BER Statistical Measurements — 249

- 8.1 Introduction — 249
- 8.2 Sampling and Statistical Histograms — 250
- 8.3 Statistical Sampling for BER — 251
- 8.4 Estimating BER, Q-factor, and SNR — 252
- 8.5 The BER Circuit — 254
- 8.6 Performance of the BER Circuit — 255
- 8.7 Other Performance Metrics — 255
- References — 256
- Standards — 258

9 Error Detection and Correction Codes — 259

- 9.1 Introduction — 259
- 9.2 Code Interleaving — 260
- 9.3 Error Detection and Correction Strategies — 260

	9.3.1	Parity	260
	9.3.2	Hamming Codes	263
	9.3.3	Convolutional Codes	266
	9.3.4	Block Codes	266
	9.3.5	Cyclic Codes	269
	9.3.6	Binary BCH Codes	270
	9.3.7	Reed–Solomon Codes and FEC	270
9.4	Summary		272
	References		273
	Standards		274

Appendix A: Statistical Noise in Communications — 277

Appendix B: Exercises and Simulation Examples Contained in the CD-ROM — 281

Index — 285

Preface

The key objectives in telecommunications—wired, wireless, or optical—are safe and cost-effective transportation and prompt delivery of the client's undamaged data. How this is done, and what technology is used to do it, is of less concern to the paying customer and of more concern to us, the telecommunications technologists. As an example, when we ship a box with some goods in it, we never ask about the specifications of the transportation equipment. We ask *when* it will be delivered and *how much* it costs, with assurance that it will be delivered undamaged.

Having said that, in megatransportation modern telecommunications systems that transport aggregate information at terabits per second, the safe and intact delivery of client's data must be guaranteed. However, unlike boxes that can be seen and touched, data are flowing at the speed of light and in massive quantities. Therefore, corrective mechanisms must be built in at the receiver end that verify the proper delivery of data at an unprecedented quality level—one failure or less in 100,000,000,000,000 bits delivered.

DWDM (Dense Wavelength Division Multiplexing), like many differently colored threads brought together to a thin rope, is a relatively new technology that enables bandwidth scalability to levels not possible before. The receiving end performs the reverse function; it unravels the rope to its constituent colored threads. Each of these "threads" is modulated to bit rates up to 40 GHz, and, thus, the potential aggregate bandwidth in a single strand of fiber is currently in excess of peta-bits per second* (1 Pbit/s = 1,000,000,000,000,000 bits) and at distances over hundreds of kilometers. As such, the sophistication of the data path as well as of the receiving end cannot be overemphasized.

An information channel that is realized with a modulated optical frequency or wavelength is termed an *optical channel*. However, as many optical channels travel in a fiber strand, many interesting phenomena take place. Light interacts with mat-

---

*Currently, terabits per second is reality. However, if the continuous spectrum from under 1300 to over 1600nm is deployed, and the channel separation is reduced, theoretically petabits per second will become a reality of the future.

ter, which interacts with light, and, as a result, although the transmitted signal was of high quality, the received signal may have been contaminated. Therefore, the amount of signal contamination needs to be estimated, monitored, and detected at the receiver so that the actual signal performance can be compared with the expected (one out of 10^{-12}) and, if it does not meet this criterion, then some remedial action must be taken based on recovery and protection strategies that are built into the system and network.

This is the second book on performance of optical channels, systems, and networks. The first book was *Fault Detectability in DWDM: Towards Signal Quality and System Robustness* (IEEE Press, 2001). This is also the fifth book on DWDM. The first book, *DWDM Networks, Components and Technology* (Wiley/IEEE 2003) provides a comprehensive treatment, with focus on the DWDM networks and how DWDM technology is employed in advanced optical systems and networks. *Introduction to DWDM Technology: Data in a Rainbow* (IEEE Press, 2000) provides an insight into the working of optical technology and an introduction to DWDM systems and networks. *Fault Detectability in DWDM* provides a treatise on fault mechanisms of DWDM components, systems, and networks and how they correlate and are detected. *Next Generation SONET/SDH: Voice and Data* (Wiley/IEEE 2004) provides a description of the next generation DWDM-based optical network and the protocols that make possible voice and data convergence over the same optical network. *Understanding SONET/SDH and ATM: Communications Networks for the Next Millennium* (IEEE Press, 1999) provides a description of the legacy SONET/SDH and ATM networks and protocols.

The objective of this book goes beyond describing optical components and their parameters, systems, and networks. The main objectives are to describe sources that affect the quality of optical signal and to provide the theoretical foundation on which the signal quality at the receiver and the performance of the optical channel are estimated, monitored, and detected. This book treats optical channels as memoryless and the signals as modulated with the most traditional on-off technique; channels with memory or multilevel modulation that are applicable to other transmission media are the subject of other textbooks and also of current research. With this objective in mind, this book reviews the fundamentals of optical communications, including modulation, the fiber as an optical transmission medium, the receiver and transmitter, jitter, and wander. It discusses factors affecting the signal quality and sources of optical noise and jitter, and how they affect the optical signal and the optical signal to noise ratio (OSNR). It clarifies the meaning of errored bits and defines bit error ratio and rate, and optical bit error rate (OBER). It discusses noise sources at the receiver, it provides a probabilistic and statistical analysis of errored bits, and links BER with SNR. It describes the eye diagram as a visualizing tool that quantifies the quality of the received signal, describes eye-diagram statistical sampling, and how BER, Q-factor, and SNR are estimated from it. It also presents a cost-efficient method for the automatic estimation of BER, Q-factor, and SNR using integrated circuitry at the receiver. This book also reviews forward error correction coding (FEC) methods, and how FEC and the estimation methodology can work together to achieve better performance.

Throughout this book there are numerical examples for pedagogical purposes. However, the real value of examples cannot be properly appreciated unless one is in a position to visualize the impact of changing parameters on the quality of the signal and quality of optical transmission. This book bridges this void by including a CD-ROM that contains ten real-life exercises that can be simulated by the reader (user, in this case) interactively. These problems are selected to cover a wide spectrum of link-layer problems, from simple attenuation to dispersion and even forward error correction. The objective of this CD-ROM and the ten problems that are simulated are described in Appendix B.

As in the previous books, it is my hope that this book will excite many communications engineers, system designers, and network architects and will stimulate many questions relevant to optical communications from DWDM network and engineering technologists as well as researchers wishing to go a step further into the interesting field of channel characterization and signal quality assessment. It is my hope that this excitation and stimulation will culminate in the design of more robust, efficient, and cost-effective optical systems and networks. I wish you happy and easy reading.

STAMATIOS V. KARTALOPOULOS, PH.D.

Acknowledgments

I thank my partner in life, Anita, and our son William and daughter Stephanie, for their consistent patience and encouragement. I also thank my publisher's staff for collaboration, enthusiasm, and project management; the anonymous reviewers for critical and constructive criticism; and all those who worked diligently on all phases of this book's production.

Constants, Conversions, and Useful Formulae

LIST OF PHYSICAL CONSTANTS

c	velocity of light in free space	$2.99792458 \times 10^{-8}$ m/sec
e	electron charge	1.60218×10^{-19} Coulomb
m_e	mass of electron	9.1085×10^{-28} gram
m_p	mass of proton	1.67243×10^{-24} gram
m_n	mass of neutron	1.67474×10^{-24} gram
m_E	mass of earth	5.983×10^{27} gram
E_e	rest energy of electron	0.51098 MeV
ε_0	permitivity of free space	8.8542×10^{-12} Farad/m
μ_0	permeability of free space	$4\pi \times 10^{-7}$ Henry/m
h	Planck's constant	$6.6260755 \times 10^{-34}$ Joule-sec
k	Boltzmann's constant	1.38×10^{-23} Joule/°K
N_A	Avogadro's number	6.0221×10^{23}/mol
$Z_0\ (\mu_0/\varepsilon_0)$	impedance of free space	376.731 Ohm (Ω)
g	acceleration due to gravity	9.80665 m/sec^2
		= 980.665 cm/sec^2
π	dimensionless number, pi	3.141593+
e	base of natural logarithm	2.7182818+
$n\ (SiO_2)$	refractive index for SiO_2, 0.8 wt % F	1.65
$n\ (Si)$	refractive index for Si	3.43
$n\ (Ge)$	refractive index for Ge	4.02

LIST OF CONSTANTS

Density of pure water at 3.98°C	1 g/ml
Density of air (at 0°C and 760 mmHg)	1.293×10^{-3} g/cm^3
Velocity of sound in air (at 0°C and 760 mmHg)	331.7 m/sec
Velocity of sound in water (at 20°C)	1470 m/sec

Refractive index (n) of PMMA acrylic (red)   1.489
Refractive index (n) of polystyrene (red)   1.584
dn/dT of PMMA acrylic $-8.0 \times 10^{-5}/°C$
dn/dT of polystyrene $-12.0 \times 10^{-5}/°C$
Linear expansion coefficient of PMMA acrylic $6.74 \times 10^{-5}/cm/°C$
Linear expansion coefficient of polystyrene $8.0 \times 10^{-5}/cm/°C$

USEFUL CONVERSIONS

1° of angle = 60 minutes = 3600 seconds = 1.7453 rads = 2.778×10^{-3} circumference
1 circumference = 360° = 2π
1 rad = 360°/2π = 0.1592 circumference = 57.296°
1 meter = 100 cm = 1000 mm = 10^6 μm = 39.37 in = 3.280833 ft = 1.09361 yd
1 km = 3300 ft
1 μm = 0.04 mils
1 nm = 10 angstroms (Å)
1 inch = 2.540005 cm
1 liter = 1000.027 cm^3 = 33.8147 fl oz = 0.26418 gal
1 ounce = 28.34953 grams
1 atmosphere = 1.0133 bar = 14.696 lb/in^2
0° C = 273 K = 32°F
1 Joule (abs) = 1 Newton-meter = 1 Watt-second = 1×10^{-7} ergs
        = 1 V-Coulomb = 9.4805×10^{-4} Btu
1 HP (electrical) = 746.00 W
1 HP (mechanical) = 745.70 W = 0.70696 Btu/sec
1 eV = 1.6×10^{-12} erg = 1.60218×10^{-19} Joule
1 wavelength (red) = 632.8 nm
1 wavelength (green) = 546.07 nm
1 wavelength (blue) = 488 nm
10 wavelengths (red) = 0.00025 inches

USEFUL FORMULAE

Probability:
$P(A) = 1 - P(A)$

Permutations:
N objects can be permuted in $n!$ ways

Combinations:
$C(n; r) = n!/[r!(n-r)!]$
$C(n; r) = C(n; n-r)$

OR rule:
$P(A \text{ or } B) = P(A) + P(B) - P(A \text{ and } B)$

XOR rule:
$P(A \text{ XOR } B) = P(A) + P(B)$

Conditional probability:
$P(A|B) = P(A \text{ and } B)/P(A)$
$P(B|A) = 1 - P(B|A)$
$P(A \text{ and } B) = P(A) P(A|B) = P(B) P(A|B)$

Independent events:
$P(A \text{ and } B) = P(A)P(B)$
$P(A \text{ and } B \text{ and } C) = P(A)P(B)P(C)$
$P(A|B) = P(A)$

Ordered events:
$P(A \text{ before } B) = P(A)/[P(A) + P(B)]$

Baye's theorem a sample space S is partitioned into n mutually exclusive events A_i; B is any event in S. Then for any i:
$P(A_i|B) = [P(A_i)P(B|A_i)]/[P(A_1)P(B|A_1) + P(A_2)P(B|A_2) + \ldots + P(A_n)P(B|A_n)]$

Binomial distribution (Bernoulli trials)
$P(k \text{ successes in } n \text{ trials}) = C(n; k)p^k q^{n-k}$

Geometric distribution:
$P(n) = (1-p)^{n-1}p$; n is the number of trials

Poisson distribution:
$P(k) = e^{-\lambda}\lambda^k/k!$ for $k = 0, 1, 2, \ldots$; λ is a parameter

Series:
$e^x = \Sigma x^k/k!$; sum is from $k = 0$ to infinity

Properties of logarithms
1. $\log(AB) = \log A + \log B$
2. $\log(A/B) = \log A - \log B$
3. $\log(A^N) = N \log A$
4. $\log\{N\text{root}(A)\} = (1/N)\log A$
5. $\log 10 = 1$
6. $\log 1 = 0$
7. $\log A = \log(e) \ln A$

8. $\log(e) = \log(2.71828+) = 0.434294$
9. $\ln 10 = 2.30258+$
10. $\ln 2.71828 = 1$
11. $\log A = +\log A$, where $A > 1$
12. $\log A = -\log|A|$, where $0 < A < 1$
13. $\log(A + B)$ not equal to $\log A + \log B$

Notice: This book contains device specifications for illustrative and educational purposes. As specifications are under continuous change, no responsibility is assumed for correctness. The reader is encouraged to consult the latest versions of standards as well as the most current manufacturer's data sheets.

Introduction

The telegraph was the first, albeit simple, data communications network, dating from the time copper wires were first deployed to distribute electrical power. Telephony was developed much later and its evolution paralleled that of the telegraph network. At about the end of the 1980s, telegraphy was abandoned because its business was overtaken by better technology; in fact, one of the T's in the original acronym AT&T stood for "Telegraph" and with the termination of the telegraph this acronym became a mere logo. The reason is that technology, such as transistor density, computer processing power, computer communications, the Internet, and so on, evolved at an exponential pace (Figure 1).

Computer communications sprang from a need to transmit files from one computer to another at high speed and at low cost, but not with the real-time deliverability, quality of service, and reliability of the voice network. Yet, the results were so well accepted that the computer network grew into a data network and, based on data protocols that encapsulated data and overhead in a packet, the Internet Protocol network was born. Thus, the traditional traffic pattern changed as now users were communicating over the Internet during the day and after work hours and for extended periods (Figure 2). To date, data traffic has grown to match voice traffic volume in traditional telecommunications networks (Figure 3).

Other data technologies were also developed taking advantage of the public switched telecommunications network (PSTN). Such technologies include the frame relay (FR) and the asynchronous transfer mode (ATM). As a result, the bandwidth demand increased such that the copper-based legacy network was not cost-efficient and a new optical-fiber-based network was created that promised to accommodate in a scalable manner the ever-increasing bandwidth demand (Figure 4).

To date, both synchronous and asynchronous networks are digital. Digitized voice and analog signals (video) as well as computer-generated data are partitioned in either synchronous bytes or in asynchronous ensembles of bytes that we call "packets." In either case, bytes or packets may be considered as ensembles of bits—eight for a byte and a multiplicity of eight for a packet. Thus, bytes or packets consist of bits, each bit having one of two values (hence, binary), and each value is represented by one of two symbols, a logical one (1) or a logical zero (0). In theory, these two values can be represented by any two symbols.* In practice, they are rep-

---

*Ancient Mesopotamians had developed an ingenious yet cumbersome numbering system that used two symbols (a "T"-like stroke and a "<"-like stroke on clay tablets). This, however, was based on a combined base 10 and base 60 numbering system and not a base 2 or binary one.

2 INTRODUCTION

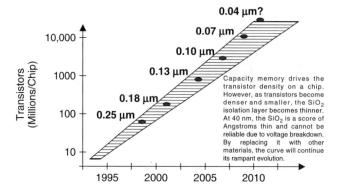

Figure 1. Evolution of transistor technology definition and density in application specific integrated circuits (ASIC).

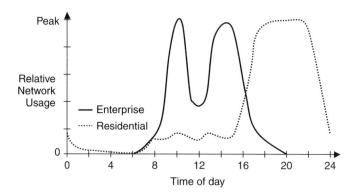

Figure 2. Traffic intensity over the span of a typical day.

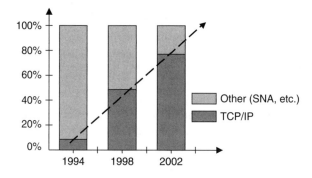

Figure 3. Growth of Internet data traffic with respect to other traffic in the telecommunications network.

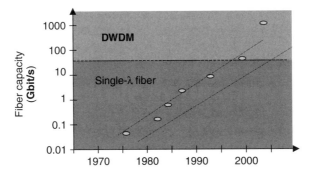

Figure 4. The fiber network enabled very high bandwidth transport. In just two decades, the transportable bandwidth per fiber exploded by a factor of 10,000!

resented by two voltage levels [0 and +V (unipolar), –V and +V (bipolar)], presence of light or lack of light, one frequency or another, one phase or another, one polarization state or another, or one spin or the opposite spin.

However, unlike bytes of digitized voice, packets are not generated in a continuous and synchronous manner. For example, when a "send file" command is issued, a string of packets is assembled and transmitted. After that, there are may be no packets to send until the next command. Thus, contrary to synchronous telephony, packets are generated sporadically. Now, if we monitor the aggregate data (packet) flow at an output, because of the sporadic or asynchronous nature of data flow, a natural bit rate is observed, which impacts how packets are generated. Thus, generated packets (TCP/IP) may have a variable length or their length may be fixed (ATM) (Figure 5). Each case has its own advantages and disadvantages, such as buffering and timing.

Transmission engineering is concerned with design issues that impact the timely transport of data at an acceptable quality. This includes a plethora of physical layer devices such as the transmitter, the receiver, the medium, amplifiers, filters, compensators, equalizers, multiplexers/demultiplexers, connectors, and others. However, components degrade or fail, affecting the quality of the signal.

Another concern of transmission engineering is link dimensioning. When bandwidth is exhausted, the links between nodes are reengineered. In this case, either the number of fibers is increased, or the number of wavelengths per fiber, or both. In either case, more components are used, which increases the probability of degradation* or failure†. Thus, if the number of fibers is increased from N to $(N+1)$, while keeping the number of wavelengths W constant, then an improvement function $F_F(N)|_W$ is defined for an acceptable grade of service, which is expressed in terms of

---

*Degradations impact the signal-to-noise ratio and, thus, the optical bit error rate, which is the subject of this book.
†The subjects of capacity exhaustion and optical fault detectability have been treated in previously published books by the same author, as indicated in the Preface.

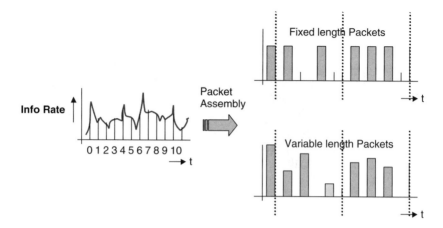

Figure 5. Based on the natural information rate, fixed-length packets are randomly generated. Variable packet length packets are more uniformly generated in time. Clearly, these two methods have different buffering and timing requirements.

a cost ratio ε and a constant γ, typically $0 < \gamma < 1$, approximated to $F_F(n)|_W = \varepsilon[1 - \gamma(1 - \varepsilon^2)]$.

Now, as the number of wavelengths per fiber is increased from W to $W + 1$, while keeping the number of fibers N constant, then an improvement function is defined, $F_W(W)|_N$ for an acceptable grade of service. In this case, the cost ratio is ε^* and the improvement function is approximated to $F_W(W)|_N = \varepsilon^*[1 - \gamma(1 - \varepsilon^{*2})]$.

As soon as path engineering is completed, the link budget is estimated. The link budget establishes the operating parameters of the link, from transmitter to receiver. This includes transmitter power and frequency; laser-beam profile specifications coupled onto the medium; medium specifications such as path loss, dispersion, optical gain, and noise; and, finally, receiver sensitivity, noise, and gain.

Link budget calculations start from the receiver and proceed toward the transmitter. Thus, the receiver signal level (RSL) is defined first to guarantee an acceptable signal quality, which is manifested by a predefined signal-to-noise ratio and bit error rate. Subsequent to this and working backward, the link loss (LL) is estimated and then the beam transmitter power coupled onto the fiber is calculated. Knowing the connector losses, and taking into account other parameters such as modulation method and bit rate, the effective generated power (EGP) of the laser is derived. Thus, the RSL is expressed as

$$RSL = EGP - LL - \Sigma \text{ (dB)}$$

where Σ denotes the lump sum of several attenuation mechanisms such as modulation method; linear and nonlinear interactions such as dispersion, polarization, and four-wave mixing; as well as noise and jitter, which all affect the quality of the signal.

The remainder of this book consists of nine chapters. Its organization starts with a general foundation in the field of optical transmission and gets progressively deeper into the subject of parameters that affect the optical signal, reaching its climax with the last three chapters. Specifically, the first chapter provides an overview of modulation and digital transmission. The second chapter focuses on the propagation of light in free space and particularly in matter. The objective of this chapter is to review fundamental linear and nonlinear phenomena, such as interference, diffraction, polarization, dispersion, four-wave mixing, and so on, that affect the quality of the propagating signal in the fiber medium. The third chapter provides an overview of the transmitter and the receiver and detection techniques, whereas the fourth chapter provides an overview of DWDM technology, devices, and networks, including optical amplification and optical spontaneous noise. The fifth chapter provides a description of noise sources that affect the quality of the optical signal. In addition, it provides a thorough overview of thermal, shot, and flicker noise. Thus, the first five chapters constitute the foundation of the main topic of this book: the optical bit error rate (BER). The sixth chapter provides a treatise in timing, jitter, and wander. The seventh chapter provides a probabilistic analysis of errored bits, and the eighth chapter provides a description of BER measurement techniques, including a novel statistical sampling method with integrated circuit. The ninth chapter provides a description of BER ramifications using error detection and correction codes. Finally, this book is enhanced with numerical exercises and a CD-ROM that includes ten interactive exercises to demonstrate how the quality of an optical signal is affected when specific parameters change. These exercises are simulated using the *OptSim*™ simulation tool from RSoft Design Group, Inc., and they enable the user to enter problem-specific parameters within the boundaries of the specific exercise.

CHAPTER 1

Principles of Modulation and Digital Transmission

1.1 DIGITAL VERSUS ANALOG

Transmission of voice over wired cables used analog electrical signals that emulated the acoustic voice signal within a frequency band that was bounded between 300 and 3400 Hz. The analog signal suffered attenuation, cross-talk, and electromagnetic interference, and it was difficult to multiplex with others. New digital techniques periodically sampled the analog signal (at 8,000 samples per second) and converted each sample to eight bits using a digital pulse-coded modulation (PCM) method (Figure 1.1). Converting the analog signal to PCM digital meant that many voice signals could be multiplexed according to an established synchronous hierarchy. Thus, data rates increased from the 64,000 bits/s (known as digital signal level 0 or DS0) to a time slot multiplexed 1.544 Mbit/s (known as DS1) and higher (Table 1.1).

In fact, up to the 1970s, long-distance "high-traffic" was at the DS3 rate. However, these were the days when cell-phones and the Internet did not exist and electrical information did not flow in large volumes. Today, the old wired network has been replaced by an optical fiber network that is capable of transporting multiple channels at an aggregate traffic of several Terabits/s per fiber; a Terabit/s is an astonishing information rate at which over a single fiber (which is thinner than the human hair) and in one second the contents of 20,000 volumes of an encyclopedia, or the entire contents of twenty movies can be transported. Currently, the maximum commercially available data rate per channel is at 40 Gbit/s (Table 1.2).

Similarly, the wireless analog mobile communications system (AMPS) is slowly being replaced by a digital system. One of the reasons for this is that the digital signal is more forgiving to noise and interference than the analog, thus improving performance, and the electronics involved with it are easier to integrate in miniaturized low-power and low-cost chips, yielding very small form-factor appliances (the current evolution of the cellular phone is a testimony to this).

However, the terms analog and digital do not only refer to the old system (analog or digital) but also to the shape of the pulse. Thus, in many respects, the digital technology at the receiver is similar to analog if one considers that digital pulses in a signal are continuously varying like a purely analog signal.

Optical Bit Error Rate. By Stamatios V. Kartalopoulos
ISBN 0-471-61545-5 © 2004 the Institute of Electrical and Electronics Engineers.

8 PRINCIPLES OF MODULATION AND DIGITAL TRANSMISSION

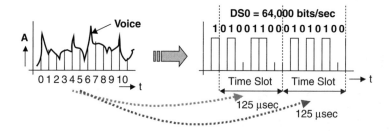

Figure 1.1. Digital telephony started with a digital signal at the lowest possible bit rate (64 kbit/s), known as DS0.

Table 1.1. Legacy rates

Facility	USA	Europe	Japan
DS0/E0	64 kbps	64 kbps	64 kbps
DS1	1,544 kbps		1,544 kbps
E1		2,048 kbps	
DS1c	3,152 kbps		3,152 kbps
DS2	6,312 kbps		6,312 kbps
E22		8,448 kbps	
			32,064 kbps
E31		34,368 kbps	
DS3	44,736 kbps		
DS3c	91,053 kbps		
			97,728 kbps
E4		139,264 kbps	
DS4	274,176 kbps		
			397.2 Mbps

Table 1.2. SONET/SDH rates

Signal Designation			
SONET	SDH	Optical	Line Rate (Mbps)
STS-1	STM-0	OC-1	51.84
STS-3	STM-1	OC-3	155.52
STS-12	STM-4	OC-12	622.08
STS-48	STM-16	OC-48	2,488.32
STS-192	STM-64	OC-192	9,953.28
STS-768	STM-254	OC-768	39,813.12

In addition, there are data rates: 10 Mbps, 100 Mbps, 1GbE, 2Gb/s, 10GbE, 40GbE(?)

In this book, by digital modulation we mean that the optical signal is binary; it is a continuous stream of two symbols, one and zero, and the two symbols are represented by an optical power pulse (logic one) or no pulse (logic zero), although other possible modulation techniques such as phase-shift keying are possible. For completeness, we review some well-known optical modulation schemes.

1.2 SPECTRUM IN OPTICAL COMMUNICATIONS

Optical communications utilize a spectrum that is in the infrared band (Figure 1.2) because the fiber medium exhibits its lowest attenuation and optimum dispersion performance in this spectrum. One should not confuse the term "optical" with "visible," as infrared is invisible to the human eye. As a consequence, laser devices are designed to generate frequencies in this (invisible) spectrum. Moreover, for interoperability purposes, the frequencies are specified to form a standard grid as recommended by ITU-T.

1.2.1 ITU-T Nominal Center Frequencies

ITU-T G.692 (October 1998) has recommended 81 optical channels in the C-band. However, these center frequencies have been extended in the L-band (longer wavelengths) as well as in bands with shorter wavelengths than the C-band. In short, starting with 1528.77 nm (or 196.10 THz) and incrementing/decrementing by 50 GHz (or 0.39 nm), a table with all center frequencies (wavelengths) is constructed. However, as fiber technology evolves, optical components improve, and WDM

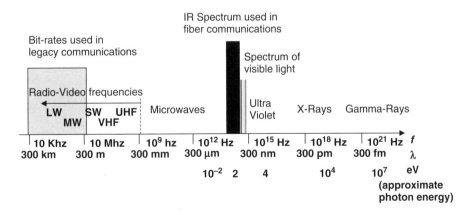

Figure 1.2. The optical spectrum used for transmission over fiber is the far infrared. The term "optical," however, should not be confused with "visible," as the far infrared is invisible to the human eye.

channel density increases when channels are incremented by 25 GHz (and perhaps by 12.5 GHz), doubling (or quadrupling) the number of channels.

1.3 LINEAR RESPONSE TO SQUARE INPUT PULSES

The response of a filter (or any system) to an input signal has been studied extensively in the literature. Here we encapsulate a brief explanation of what occurs when a signal passes through a medium. This discussion will be used in this book, particularly for amplitude modulated signals.

Consider an ideal square pulse of period T. This pulse propagates through a medium and interacts with it. To study the effect of interaction on the pulse, the medium is described by an analytical function in terms of its parameters and frequency; this is known as the transfer function $H(\omega)$. Thus, the output response of the medium to the input signal is calculated from the product of the two.

If the medium is linear, then the transfer function $H(\omega)$ has a response depending on the input signal frequency (or period) and the poles and zeroes of the transfer function, that is, the frequency values for which the numerator of $H(\omega)$ becomes zero (the zeroes), the values for which the denominator becomes zero (the poles), and for which the transfer function becomes infinite. Thus, the filter parameters, the signal amplitude, and frequency (or period of the ideal square pulse) determine the damping factor and the poles and zeroes of the transfer function, the overshoot and undershoot, the ripple, and so on. Remember that, from Fourier analysis, a square pulse of period T consists of a specific and rich frequency spectrum.

If the medium is nonlinear, then this analysis is more complex. In this case, as a rule of thumb, the transfer function is expanded in a series, the first term of which is linear and it is studied as previously explained. Subsequent terms—second degree, third degree, and so on—represent nonlinearities that are studied with complex mathematical analyses (such as Voltaire series) and approximations. However, it is very rare to have to analyze nonlinearities beyond the second or third degree. In fact, when the frequency operating point and the signal amplitude are selected carefully, in many cases even nonlinear systems may be considered good linear approximations.

If we assume that a system's response is similar to a filter, then the filter may be viewed as a frequency window, permitting certain frequencies to pass through but with a specific toll on each one. Thus, if the filter bandwidth is ω_0 (rad/s) and the input pulse has a period T, then there are three cases of interest: $B_S \gg 1/T$, $B_S = 1/T$, and $B_S \ll 1/T$, where $B_S = \omega_0/2\pi$ (Hz). Typically, B_S or ω_0 is determined at the 3 dB point (or 0.707) of the filter response.

When $B_s \gg 1/T$, the frequency window of the filter is wide enough to pass most frequencies of the pulse and with little interaction. In such cases, the output resembles the input with some added overshoot, undershoot, attenuation (or gain), and rise/fall time (t_r/t_f) as a result of parametric changes (Figure 1.3A). If rise time (t_r) is measured from the -0.09 to the 1.09 point of the output pulse, t_r is estimated to be $t_r = 0.8/B_S$, and if it is measured at the half-amplitude point of the output pulse, then t_R

= $0.5/B_S$. The rise time of the output pulse is, in general, inversely proportional to the filter bandwidth B_S. The amount of overshoot and undershoot depends on the damping factor of the (higher-order) filter.

When $B_S=1/T$, then the filter passes the fundamental frequency of the pulse with a rise time about half the pulse width (Figure 1.3B).

When $B_S \ll 1/T$, then the filter passes the fundamental frequency of the pulse but grossly attenuated (Figure 1.3C). This could be viewed as a very narrow frequency window through which a much wider frequency band tries to pass.

1.4 PRINCIPLES OF MODULATION

In general, a laser semiconductor device (or semiconductor laser) generates a continuous photonic beam. However, as in its simplest mode of operation a semiconductor laser produces a continuous wave (CW) or beam of light, this beam carries no information other than frequency or wavelength. However, when the beam is modulated, then it could carry data at a currently commercially available data rate of up to 40 Gbits/s (that is, 40,000,000,000 bits per second), and research continues to seek higher modulation rates.*

Modulation is the action of temporally altering one or more of the parameters of the photonic signal. In optical communications, such parameters are phase, frequen-

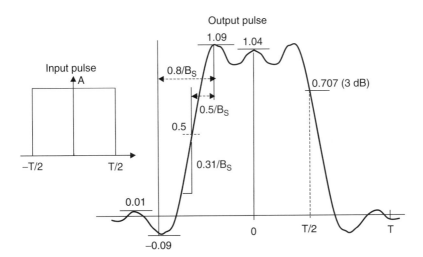

Figure 1.3. (A). Anatomy of an output pulse from an ideal square pulse at input for the case $B_S \gg 1/T$.

---

*40 Gbits/s corresponds to more than 600,000 simultaneous conversations. It also corresponds to many thousand simultaneous video channels to satisfy the video-on-demand needs of a large city. In short, such data rates satisfy all voice, data, and video needs of a sizeable modern city.

12 PRINCIPLES OF MODULATION AND DIGITAL TRANSMISSION

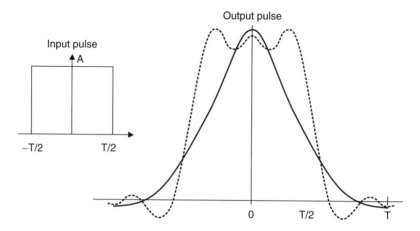

Figure 1.3. (B). Anatomy of an output pulse from an ideal square pulse at input for the case $B_S = 1/T$. For comparison, it is superimposed on the pulse from an ideal square pulse at input for the case $B_S \gg 1/T$ case (dotted line).

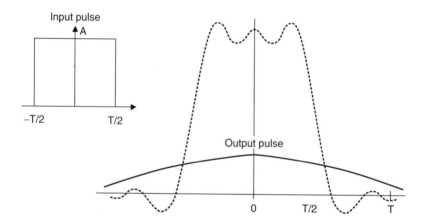

Figure 1.3. (C). Anatomy of an output pulse from an ideal square pulse at input for the case $B_S \ll 1/T$. For comparison, it is superimposed on the pulse from an ideal square pulse at input for the case $B_S \gg 1/T$ case (dotted line).

cy, polarization, and amplitude. When the phase is modulated, the method is called *phase shift keying* (PSK); when the frequency is modulated, it is called *frequency shift keying* (FSK); when the state of polarization is modulated, it is called *state-of-polarization shift keying* (SoPSK); and when the amplitude is modulated, it is called *amplitude shift keying* (ASK). The latter case includes the *intensity modulation with direct detection* (IM/DD) and the on–off keying (OOK) modulation methods.

In optical communications, the modulation method plays a key role in the:

- Optical power coupled into the fiber
- Bit-rate limits
- Transportable amount of information per channel
- Dispersion limits
- Fiber-span limit
- Linear and nonlinear contributing effects
- Overall signal-to-noise ratio and bit error rate
- Reliability of signal detection and receiver penalty

1.4.1 On–Off Keying and NRZ Versus RZ

The On–Off Keying (OOK) modulation is an amplitude-modulation (AM) method. The modulating signal closely resembles a square pulse that acts as a shutter on the laser beam, hence its name. The OOK method generates a stream of pulses that are then transmitted over the fiber. Figure 1.4 illustrates a binary pulse train, the spectrum of the modulating signal, and the spectrum of the modulated signal. When the logic "one" is lighted for the full period ($T = 1/f$), this OOK is termed *nonreturn to zero* (NRZ), and when for a fraction of the period (such as ⅓ or ½), it is termed *return to zero* (RZ). As a consequence, NRZ modulation utilizes the full period as compared with RZ, which is a fraction of it. Thus, the energy within a NRZ bit is

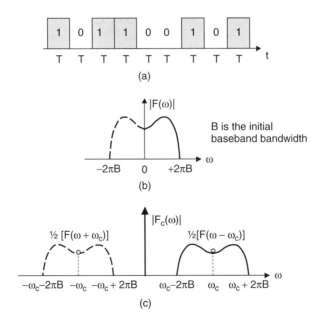

Figure 1.4. A binary OOK pulse train (A), the spectrum of the square modulating pulse (B), and the spectrum of the modulated signal (C).

much more than the energy in a RZ bit, if everything else remains the same. This implies that either the NRZ signal can propagate to longer distances than the RZ or the NRZ power level can be lowered, for the same distance.

OOK modulation can be used in both coherent and direct detection. However, coherent detection requires phase stability. As a consequence, the laser source cannot be directly modulated, as this may shift the signal phase and add chirp. To reduce this, the signal amplitude is modulated externally using a titanium-diffused $LiNbO_3$ waveguide in a Mach–Zehnder configuration or a semiconductor directional coupler based on electroabsorption multiquantum well (MQW) properties and structures. On the other hand, direct detection does not require stable phase; however, direct modulation may alter the spectral content of the source, which raises other issues.

In WDM optical communications, the carrier frequency is the electromagnetic wave in the infrared wavelength range (800 to 1620 nm). In single-wavelength transmission, selected wavelengths were used; 880 nm was used in multimode short-fiber applications and 1310 or 1550 nm in single-mode long-haul fiber applications.

The OOK amplitude modulated carrier is expressed as

$$f_c(t) = A \cdot a_m(t) \cos \omega_c t$$

where A is the amplitude of the OOK signal and $a_m(t)$ is 0 or 1 during an interval T. The Fourier transform of the OOK signal, using the frequency-shifting theorem, is expressed as

$$F_c(\omega) = \frac{A}{2}[F(\omega - \omega_c) + F(\omega + \omega_c)]$$

Figure 1.4 provides a pictorial representation of the above mathematical analysis. Notice that for an initial baseband bandwidth B hertz (or $2\pi B$ rad/s), the modulated bandwidth is twice that, $\pm B$ hertz about the carrier, or $2B$ hertz. This is the double-sideband (DSB) case. In other words, if the carrier spectrum is sliced into frequency channels, each slice must be at least $2B$ wide (in practice, it must be much wider to accommodate other effects). If one of the sidebands is rejected and the other is used, then this is called the vestigial-sideband (VSB) case.

The OOK modulating method is also known as pulse amplitude modulation (PAM). In optical communications, we limit ourselves to the practical case of binary signals (two symbols, 0 and 1). However, studies have also analyzed M-ary signals (M symbols). In this case, the M-ary modulated signal $S_m(t)$ is mathematically expressed as

$$S_m(t) = A[(M-1) - 2a_m]p_T(t)$$

where A is the amplitude of the signal, a_m is the binary sequence, and $p_T(t)$ is the pulse shape. This expression yields $M/2$ pairs of antipodal signals, $\pm A$, $\pm 3A$, $\pm(M-1)A$. In optical communications, this method has not been used in commercial terrestrial applications.

1.4.2 A Case of Amplitude Modulation

An analysis of amplitude (intensity) modulation of an optical channel yields some interesting results that explain certain degradations observed during experimentation with very fast bit rates (>10 Gbit/s). Here, we present a simple modulation case to identify these results.

Consider a monochromatic signal, ω_c, which is amplitude modulated by a function $g(t)$, and described by

$$m(t) = g(t) \cos \omega_c t$$

The amplitude modulating function $g(t)$ is described in general by

$$g(t) = [g_0 + mv(t)]$$

and thus,

$$m(t) = [g_0 + mv(t)] \cos \omega_c t$$

where m is the modulation index (equal to 1 for 100% modulation); g_0 is a DC component, which for simplicity can be set to 1; and $v(t)$ is the modulating function.

Consider the simple case where $v(t) = g_m \cos \omega_m t$. Based on this, the modulated signal $m(t)$ may be trigonometrically expanded:

$$m(t) = g_0 \cos \omega_c t + (m/2)[g_m \cos(\omega_c - \omega_m)t] + (m/2)[g_m \cos(\omega_c + \omega_m)t]$$

This expansion yields three terms: the main frequency (first term) and two sidebands, each ω_m far from ω_c (Figure 1.4). In more complex cases, trigonometric expansion becomes more elaborate.

In a (complex) vectorial representation, the latter relationship is written as

$$m(t) = Re[e^{j\omega_c t} + (m/2)e^{j(\omega_c - \omega_m)t} + (m/2)e^{j(\omega_c - \omega_m)t}]$$

where Re denotes the real part of the complex exponential notation. If the exponential terms are viewed as phasors, then $m(t)$ consists of three terms: one stationary and two counterrotating terms: the sum of which yields the modulation signal. If the amplitude of the carrier is unity, then each sideband has a power of $m2/4$, and both $m2/2$.

Based on this analysis and under certain worst-case conditions, an interesting degradation may take place. If the lower sideband frequency ($\omega_c - \omega_m$) is shifted clockwise by q degrees and the upper sideband ($\omega_c + \omega_m$) clockwise by $180 - q$ degrees, then the resultant vector represents a phase-modulated wave, the amplitude modulation of which is largely cancelled, or the modulation index becomes zero. Clearly, in cases in between, the degradation of the modulation index is partial.

16 PRINCIPLES OF MODULATION AND DIGITAL TRANSMISSION

In optical communications, the two sidebands represent different wavelengths, one at λ_1 corresponding to $(\omega_c - \omega_m)$ and another at λ_2 corresponding to $(\omega_c + \omega_m)$ which, because of dispersion phenomena, travel at different speeds and, thus, at different phases. Consequently, on–off keying (OOK) modulation, under certain conditions, is expected to trigger certain interesting phenomena such as intersymbol interference, modulation instability, and others.

1.4.3 Phase-Shift Keying

This method modulates a light beam (the carrier) by changing the phase of the carrier (by 180 degrees) at the transition from logic one to logic zero and vice versa; that is, it shifts the phase by 180 degrees while the frequency and amplitude of the signal remain constant for all bits, thus appearing as a continuous light wave (Figure 1.5). For multilevel PSK, the change may be in increments of 45 degrees (8 levels). As a consequence, PSK requires a coherent carrier.

PSK is implemented by passing the light beam through a device that operates on the principle that, when a voltage is applied to it, its refractive index changes; this is known as *electrorefraction modulation* (ERM). Such devices are made with electrooptic crystals such as $LiNbO_3$, with proper orientation. The phase-shift keying signal is described as

$$f_c(t) = \pm \cos \omega_c(t + \tau) \qquad -T/2 \leqq t \leqq T/2$$

The phase difference is expressed by

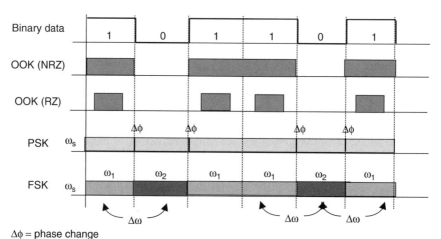

Figure 1.5. Optical modulation methods.

$$\delta\phi = (2\pi/\lambda)(\delta n)L_m$$

where the index change δn is proportional to applied voltage, V, and L_m is the length over which the index changes by the applied voltage.

1.4.4 Frequency-Shift Keying

The binary signal can be conveyed by changing the frequency ω of the carrier at the transitions between logic "zero" and logic "one"; that is, it shifts the frequency while the amplitude of the signal remains constant for bits. At the transitions, the frequency changes by Δf—$f + \Delta f$ for logic "1" and $f - \Delta f$ for logic "0" (Figure 1.5). Typical frequency changes are about 1.

The total spectral bandwidth of an FSK signal is approximated to $2\Delta f + 2B$, where B is the bit rate and Δf the frequency deviation.

- When the deviation is large, $\Delta f \gg B$, the spectral bandwidth approaches $2\Delta f$, and this case is known as *wideband FSK*.
- When the deviation is narrow, $\Delta f \ll B$, the spectral bandwidth approaches $2B$, and this case is known as *narrowband FSK*.

A *frequency modulation index* (FMI), defined by $\Delta f/B = \beta_{FM}$, distinguishes the two cases: wideband FSK has an FMI of $\beta_{FM} \gg 1$ and narrowband FSK has an FMI of $\beta_{FM} \ll 1$.

FSK is achieved with electroacoustic Bragg modulators or with DFB semiconductor lasers that shift their operating frequency by 1 when the operating current changes by a mere 1 mA. Thus, DFB semiconductor lasers make very good and fast coherent FSK sources with high modulation efficiency. Figure 1.5 summarizes all aforementioned shift-keying modulation methods.

1.4.5 Duo-Binary Modulation

The duo-binary modulation case is an amplitude modulation case but in this case the bitstream is manipulated to reduce the bandwidth and certain phenomena that lead to intersymbol interference (ISI).

The duo-binary method was first applied to electrical digital signals. It combines two successive binary pulses in the digital stream to form a multilevel electrical signal. Then, at the transmitting end, if the bit stream is x_k (this also is the input to the encoder), the output of the duo-binary encoder is $y_k = x_k + x_{k-1}$, which is derived as shown in Figure 1.6:

$$x_k: \quad -1 \quad -1 \quad 1 \quad -1 \quad 1 \quad 1 \quad 1 \quad -1 \quad -1 \quad -1 \quad 1 \quad -1$$
$$y_k: \quad \quad -2 \quad 0 \quad 0 \quad 0 \quad 2 \quad 2 \quad 0 \quad -2 \quad -2 \quad 0 \quad 0$$

Based on this, duo-binary encoding trades bandwidth for effective signal strength

18 PRINCIPLES OF MODULATION AND DIGITAL TRANSMISSION

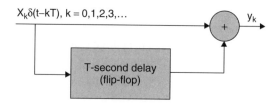

Figure 1.6. A binary bit stream mathematically represented by a sequence of narrow pulses $X_k\delta(t - kT)$ is summed with its delayed bit to yield a duo-binary signal y_k. This output can sustain filtering by a low-pass filter without loss of information. T is the period of the bit rate and δ is the Dirac function.

and, thus, it affords bandwidth reduction (and ISI). It is also possible to apply precoding before coding to further improve the method and simplify the receiver.

At the receiving end, the reverse process takes place: the received duo-binary signal y_k is decoded to retrieve the original digital signal x_k. That is, y_k is the input to the decoder and x_k is the output:

y_k: −2   0   0   0   2   2   0   −2   −2   0   0
x_k: −1 −1   1 −1   1   1   1 −1 −1 −1   1 −1

Notice that in this scheme there are three levels (−2, 0, and +2) and that the absolute maximum level is |2|. As a result, a bit error at the receiver, although it may initially propagate, it is eventually is detected, as shown in the following examples.

Example #1
In this example, the third received bit is in error (underlined) and it causes erroneous outputs at the decoder, starting with the sixth output bit (underlined):

y_k: −2   0   <u>2</u>   0   2   2   0   −2   −2   0
x_k: −1 −1   1   1 −1   <u>1</u>   1 −1 −1 −1   1 −1

Example #2
In this example, an error in the sixth received bit results in an impossibility at the ninth generated bit (X):

y_k: −2   0   0   0   2   0   0   −2   −2   0   0
x_k: −1 −1   1 −1   1   1 −1   1   <u>X</u>

Example #3
In this example, as in Example #1, an error in the third received bit will be detected by the sixth generated bit (X):

y_k: −2 0 2 0 2 2 0 −2 −2 0 0
x_k: −1 −1 1 1 −1 X

In the above examples, we have used the two symbols −1 and +1. The method can be extended to optical binary using the symbols 0 and 1 (since negative light, which would be represented by −1, does not exist). Then, at the transmitter:

x_k: 0 0 1 0 1 1 1 0 0 1 0
y_k: 0 1 1 1 2 2 1 0 0 1 1

And at the receiver:

y_k: 0 1 1 1 2 2 1 0 0 1 1
x_k: 0 0 1 0 1 1 1 0 0 1 0

In optical communications, the duo-binary levels may represent either optical power in three different power levels (as the last example), one of three frequencies, one of three phases, or one of three polarization states. Clearly, power intensity suffers attenuation and, with additive noise, it is questionable if this method can be used for long fiber lengths with optical amplifiers, although it may be suitable in certain short-haul and fiber-to-home applications. Three frequencies per channel implies that for spectral efficiency a channel must be sliced into three subspectral ranges, which may be possible in CWDM but perhaps not in DWDM applications. The third solution is more plausible, as it may be applicable to both CWDM and DWDM. In this case, the three symbols (−2, 0, +2) or (0, 1, 2) may represent a phase shift of −90, 0, and +90 degrees. Thus, in optical communications, this duo-binary version corresponds to a phase-shift-keying (PSK) method (Figure 1.7) that may be preferred in ultrahigh bit rate applications.

1.4.6 State-of-Polarization Shift Keying

This method varies the polarization of light between two orthogonal states, P^T and P^{II}. However, it assumes that over the optical path, there is no strong birefringence

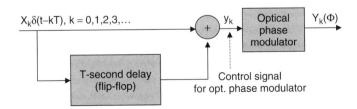

Figure 1.7. One implementation of duo-binary modulation. The y_k bit stream $X_k\delta(t - kT)$ provides the control signal to an optical-phase modulator. When the amplitude is 1, the phase is 0; when it is 0, the phase is −90; and when it is 2, the phase is +90 degrees.

or polarizing devices that may alter or shift one of the polarization states. This is an assumption, as the fiber itself exhibits a degree of birefringence B that is a measure of the index variation in the two principle axes:

$$B = |n_x - n_y|$$

Although researchers have demonstrated that the SoPSK method can be used in multilevel (two, four, many) applications, thus compressing the amount of data over one carrier, this modulation method has not been used commercially yet, because the polarization dispersion mode (PMD) is a limiting factor.

1.4.7 Comparison between OOK and PSK

The OOK modulation method provides the flexibility to modulate pulses as wide as the period, thus known as nonreturn to zero (NRZ). It also provides the flexibility to modulate pulses narrower than the period, known as return to zero (RZ).

OOK is a modulating method that affects the optical intensity of the signal. However, abrupt transitions of the modulating signal (ultrashort t_r and t_f) affects the nonlinear behavior of the modulator, yielding parasitic effects (chirp, noise, jitter). Therefore, at the receiver, the detection of the received symbols (one or zero) may be direct.

Conversely, the PSK method is phase sensitive and as the signal propagates through optical components (such as interferometric filters) they may affect the phase of the signal and degrade the phase-shift quality, which may affect the symbol (one or zero) detection at the receiver. However, the phase difference in PSK is the first step in a modulator, and at the receiver it is relatively easy to detect using interferometric methods.

1.5 MODULATOR TYPES

Optical semiconductor modulators can be combined with laser devices. Among them, the lithium niobate (LiNbO$_3$) Mach–Zehnder (M-Z), indium phosphide (InP) electroabsorption, electroabsorption multiquantum well (MQW), and electrorefraction modulators are the most interesting.

The *Mach–Zehnder* (MZ) modulator consists of a Y-splitter junction, one or two phase modulators, two waveguides, and a Y-combiner junction. The incoming optical power is split by the first Y-junction into two equal paths. The application of a field (voltage) on one of the paths causes the refractive index to change. When this is done in a controllable manner, then a 180 degree shift in the relative phase between the two paths is accomplished. Based on this, light from the two segments interferes at the recombining Y-junction destructively or constructively, and the continuous signal is modulated. Destructive interference produces a logic "zero" and constructive interference a logic "one." The modulation depth is directly related to the extinction ratio of the modulator, or the optical power of logic "one" over the optical power of logic "zero". Similarly, the quadrature bias of the modulator is also directly related to its dynamic range. Additionally, the extinction ratio of the modulator is closely linked with the BER of the signal.

The *electro-absorption* (EA) modulator is an on–off optical device made with InGaAsP. The percent of absorbed photons depends on the strength of the applied field. EAs have an almost logarithmic attenuation of optical power that depends on a reverse voltage applied to them, as opposed to laser devices that require forward bias. With no voltage applied, EA modulators are transparent and when a voltage is applied, they absorb light at the laser wavelength they are designed for. Thus, they cause a modulation depth in excess of 45 dB. They are fast devices and, therefore, they modulate a continuous beam to short optical pulses at bit rates in excess of 40 Gb/s. However, the integrated structure should be electrically isolated from the laser device to avoid parasitic influences.

Electroabsorption multiquantum well (EA-MQW) modulators are directional couplers with light-absorption properties that are based on semiconductor structures known as multiquantum wells. In this case, light is absorbed based on the voltage applied. MQWs act as fast shutters and may be integrated with DFB lasers.

Electrorefraction modulators (ER) are based on the Pockels effect and directly control the phase of an optical wave upon the application of a voltage. These modulators consist of a Ti-diffused waveguide in a $LiNbO_3$ substrate. Strip electrodes are deposited at either side of the waveguide to apply a transverse field across the waveguide. The applied voltage for a 180 degree shift depends on the length of the waveguide immersed in the field and on the plane orientation with respect to the field and the optical propagation the $LiNbO_3$ crystal is cut for.

The above modulators have a transfer function that describes the optical output power versus applied field or voltage, based on which a bias for optimum modulation is determined and must be maintained.

1.6 PRINCIPLES OF DECODING

Optical decoding entails detecting the optical signal, converting it to electrical impulses, and retrieving binary coded information (or demodulate) from the received modulated lightwave, based on one of the modulation methods described earlier:

- Detect optical amplitude level if amplitude shift keying (ASK or OOK) is used
- Detect phase change (from 0° to 180°) if binary PSK is used
- Detect frequency change (from $\omega - \Delta\omega$ to $\omega + \Delta\omega$) if FSK is used
- Detect state of polarization, P^T or P^{II}, and change if SoPSK is used

A metric of a good decoder at the receiver is when the uncertainty of the received bits (1 or 0) is less than 1 bit per second per Hertz (< 1 bit/s-Hz); bit uncertainty may result in errored bits. At 10 Gbit/s, this metric translates to less than one errored bit in 10,000,000,000 bits, or 10^{-10}. Optical communication systems are designed with a single-bit uncertainty (specified in ITU-T standards) of less than 10^{-12}. Thus, whatever decoding method is used, it clearly impacts the decoder (receiver and demodulator) design and its accuracy.

Coherent heterodyne and *homodyne* detection techniques were initially developed for radio communications. In optical transmission, the term "coherent" indicates that a light source in the vicinity of the transmitted source is used as the local oscillator at the receiver. The local oscillator is expressed as

$$C(t) = 2\cos(2\pi f_C t + \Theta)$$

where f_C and Θ are the expected local oscillator frequency and phase with respect to the incoming signal.

The aforementioned expression assumes an ideal monochromatic frequency f_C. However, in reality this is not the case as both the incoming optical signal and the optical oscillator are laser sources, presumably of the same wavelength. Therefore, the local oscillator must have a narrow spectral width that is comparable to the source (Figure 1.8). In addition, the local oscillator must have low-noise characteristics, otherwise the spontaneously emitted light adds to noise and the method is not practical. Therefore, the amplitude of the local oscillator in coherent receiver design is important. In IM/DD, the incoming signal is directly coupled into the detector, eliminating the coupler and the local oscillator.

Optical coherent methods with a stable coherent reference improve receiver sensitivity by ~20 dB, allowing longer fibers to be used (by an additional 100 km at 1.55 μm). Although IM/DD detection and demodulation requires channel spacing on the order of 100 GHz, coherent techniques support spacing as small as 1–10 GHz, although currently they are not as popular.

OOK RZ and NRZ demodulators detect directly incident photons. The temporal density of photon fluctuation generates a temporal electrical fluctuation with simi-

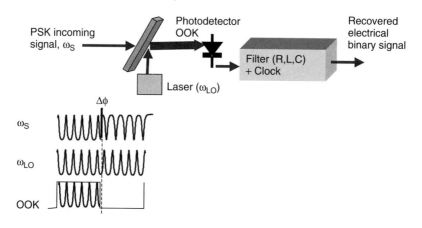

Figure 1.8. Principles of a homodyne PSK demodulator.

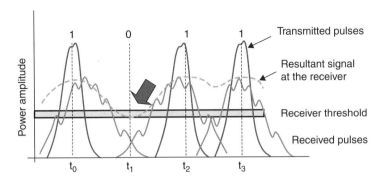

Figure 1.9. Compound effect of loss, dispersion, noise, and jitter on digital optical signal at the receiver. The arrow shows the area of uncertainty and possible bit error.

lar amplitude plus some electrical noise added by the photodetector and receiver electronics. Then, the electrical fluctuation is low-pass filtered to remove high-frequency noise. The filtered signals are then sampled at an expected bit rate at a periodic instant and a threshold level that minimizes jitter and signal-level uncertainty. Thus, the density fluctuation of incident photons is interpreted as electrical logic "1" when it is above the threshold level or as logic "0" when it is below it. However, there are instances when the incident amplitude is ambiguous due to excessive attenuation, dispersion, noise, and jitter and an erroneous symbol (1/0 instead of 0/1) may be produced (Figure 1.9).

It should be noticed that a OOK NRZ signal provides photons for the full duration of the bit period, whereas a RZ signal does so for a percentage of the period (Figure 1.10), such as 33%, 40%, or 50%. The NRZ or RZ modulation, and the percentage is particularly important in ultrahigh bit rates such as 10 or 40. For example, a 50% OOK 40 signal has logic "1" illuminated for 12.5 ps, whereas a 33% is

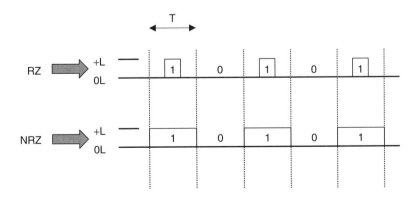

Figure 1.10. OOK RZ and NRZ optical coding.

24 PRINCIPLES OF MODULATION AND DIGITAL TRANSMISSION

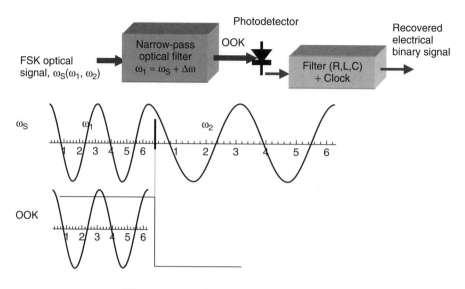

Figure 1.11. A simplified FSK demodulator.

illuminated for 8.25ps. If all things are equal, this reduction is significant in the amount of received power, and, thus, in the received bit error rate. However, if the path is engineered correctly and the RZ peak power is higher, then RZ provides better noise isolation, so RZ improves the overall optical signal-to-noise ratio.

PSK demodulation is based on coherent detection. That is, in addition to the received optical signal, one or two local oscillators (optical frequencies) are required to interferometrically interact with the received optical signal and convert it to an amplitude modulated (AM) signal. That is, the received PSK signal, ω_S, is mixed coherently with a locally generated laser light, ω_{LO}, and since both are of the same frequency, they interact interferometrically. When both frequencies are in phase, there is constructive interference, and when they are not, destructive, and, thus (ideally), an OOK signal is generated. Since the accuracy of this method depends on the phase variation of the signal, phase stability and low noise are very critical.

In FSK demodulation, the optical signal, ωS, is passed through a narrow-band optical filter tuned to pass the frequency $\omega_1 = \omega + \Delta\omega$ and reject the frequency ω_2 (Figure 1.11). Thus, the outcome is equivalent to an OOK modulated signal. Since the accuracy of this method depends on the frequency variation of the signal, no frequency shift (high-frequency stability) and low optical noise are very critical.

REFERENCES

1. M. Schwartz, *Information Transmission, Modulation, and Noise,* 3rd ed., McGraw-Hill, New York, 1980.

2. M.C. Jeruchim, P. Malabian, and K.S. Shanmugan, *Simulation of Communications Systems,* Plenum Press, New York, 1992.
3. J.C. Bellamy, *Digital Telephony,* 3rd ed., Wiley, New York, 2000.
4. R.E. Matick, *Transmission Lines for Digital and Communication Networks,* IEEE Press, New York, 1995.
5. J.C. Palais, *Fiber Optic Communications,* 3rd ed., Prentice-Hall, Englewood Cliffs, NJ, 1992.
6. I.P. Kaminow, and T.L. Koch (Eds.), *Optical Fiber Communications IIIA and* Optical Fiber Communications IIIB, Academic Press, San Diego, 1997.
7. S.V. Kartalopoulos, *DWDM: Networks, Devices and Technology,* Wiley/IEEE, Hoboken, NJ, 2002.
8. S.V. Kartalopoulos, *Introduction to DWDM Technology: Data in a Rainbow,* Wiley/IEEE, New York, 2000.
9. S.V. Kartalopoulos, *Fault Detectability in DWDM: Toward Higher Signal Quality and System Reliability,* IEEE Press, Piscataway, NJ, 2001.
10. S.V. Kartalopoulos, *Understanding SONET/SDH and ATM Networks,* IEEE Press, Piscataway, NJ, 1999.
11. C.E. Shannon, "A Mathematical Theory of Communication," *Bell System Technical Journal,* pp. 379–423, 623–656, 1948.
12. J. Tang, "The Shannon Channel Capacity of Dispersion-Free Nonlinear Optical Fiber Transmission," *Journal of Lightwave Technology, 19,* 8, 1104–1109, 2001.
13. R.A. Linke, "Optical Heterodyne Communications Systems," *IEEE Communications Magazine,* Oct. 1989, pp. 36–41.
14. R.A. Linke, and A.H. Gnauck, "High-Capacity Coherent Lightwave Systems," *Journal of Lightwave Technology, 6,* 11, 1750–1769, 1988.
15. E.A. De Souza, et al., "Wavelength-Division Multiplexing with Femtosecond Pulses," *Optics Letters, 20,* 10, 1166–1168, 1995.
16. R.E. Slusher, and B. Yurke, "Squeezed Light for Coherent Communications," *Journal of Lightwave Technology, vol. 8,* 3, 466–477, 1990.
17. A. Lender, "Correlative Level Coding for Binary Data Transmission," *IEEE Spectrum, 3,* 2, 104–115, February 1966.
18. J.A.C. Bingham, "Multicarrier Modulation for Data Transmission: An Idea Whose Time Has Come," *IEEE Communications Magazine, 28,* 5, 5–14, May 1990.

STANDARDS

1. ANSI/IEEE 812-1984, "Definition of Terms Relating to Fiber Optics," 1984.
2. ITU-T Recommendation G.701, "Vocabulary of Digital Transmission and Multiplexing, and Pulse Code Modulation (PCM) Terms," 1993.
3. ITU-T Recommendation G.702, "Digital Hierarchy Bit Rates," 1988.
4. ITU-T Recommendation G.704, "Synchronous Frame Structures Used at 1544, 6312, 2048, 8488 and 44736 Kbps Hierarchical Levels," 1995.
5. Telcordia, TR-NWT-499, "Transport Systems Generic Requirements (TSGR): Common Requirements," issue 5, Dec. 1993.

CHAPTER 2

Optical Propagation

2.1 INTRODUCTION

The study of optical signals propagating in matter requires a good understanding of the physics of light and the properties of matter. How is light generated? What are its nature and its properties? How does light interact with matter? How is the optical signal affected by this interaction? In this chapter, we address these questions to help in the understanding of light–matter interaction and how this affects signal quality.

Over the years, the nature of light has been explained based on the semiclassical theory of the atom. Excited electrons rotate around the nucleus of the atom in a higher-energy elliptical orbit, E_i, similar to how planets revolve around the sun. Being at a higher energy, the electron rotates at a high rpm. Now, when the electron is suddenly forced to a lower rpm, and thus to a lower rotational energy, an amount of energy is released. That is, when the electron transits from a high-energy orbit to a lower one, energy equal to the difference between the two energy levels is released, $E_{released} = E_i - E_j$. However, since electron orbits are quantized (they are not continuously distributed around the nucleus), so the amount of energy released is quantized. As a consequence, specific atoms release specific amounts of energy, as electrons transit from specific high-energy levels to specific low-energy levels, which are characteristic for each element of the periodic table. These energy quanta manifest themselves as photons if their energy is a multiple of Planck's constant, h.

Up to this point, this description neither tells us much about the characteristics of the released photon, namely frequency, nor its dual electromagnetic and particle-like nature. To further explore the nature of the photon, consider that an excited electron moves in some elliptical orbit around the nucleus at some angular frequency or rpm. Thus, the energy of the electron may be expressed as a function of its angular frequency ω_1, $E(\omega_1)$. Since the electron has a charge and it is in continuous motion, it represents a current creating a magnetic field around its elliptical orbit. This magnetic field may be viewed (for simplicity) as an evolving toroidal and an electric field perpendicular to it and the orbit of the electron (Figure 2.1).

Now, as the electron "moves" from an electromagnetic field rotating at ω_1 to another with lower energy and angular frequency ω_2, the difference between the two energies $E(\omega_1) - E(\omega_2)$ corresponds to the released energy. If each magnetic and electric field of each level is expressed as a complex function of ω, such as $F(\cos \omega$

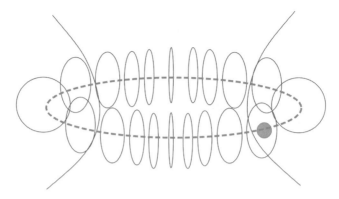

Figure 2.1. An electron (dot) moves on an elliptical trajectory (dotted line) and generates an evolving magnetic field (circles) in a toroidal shape and at a frequency commensurate with its angular velocity ω. The direction of the magnetic field depends on the direction of motion of the electron. If clockwise, the magnetic field is also clockwise.

$+j \sin \omega$), then the difference of the two levels provides a cosine term, the argument of which is $\omega_1 - \omega_2 = \omega_3$. Thus, the released energy is an electromagnetic wavelet at frequency ω_3, which at the moment of its release is perpendicular to the orbital plane of the electron. However, the released wavelet is no longer constrained to move on a toroidal-shaped orbit, but escapes the atom's influence, unraveling the toroidal shape and propagating unbounded in a straight path in a corkscrew-like manner (Figure 2.2). In this way, the magnetic and electric fields still remain orthogonal and revolve around each other. Thus, this model explains the quantum-mechanical electromagnetic nature of light, its frequency and unique "color," and its propagation characteristics in free space.

It turns out that the energy of the released electromagnetic wavelet, the photon, is directly proportional to its frequency, with a constant of proportionality, Planck's constant, $E = h\nu$. Thus, one would be tempted to say that created photons move like tiny electromagnetic "fuzzy" balls with total energy given by $E = h\nu$ and moving in free space at the speed of light defined in Einstein's energy relationship $E = mc^2$. In short, photons have particle properties and thus momentum that exerts pressure on a target, which has been verified by many experiments (Figure 2.3), and it is routinely experienced by satellites when they enter the path of solar light. However, the philosophical search for the essence of light is intriguing, and still continues. A theory that attempts to bridge the classical with quantum mechanics has been in active development for the last 20 years and is known as the string theory (or "the theory of everything"). According to it, everything consists of subquantum-mechanical, multidimensional, oscillating strings, a bunch of which may define gravitons, nuclear particles, electrons, photons, and so on. However, although this theory (or theories, as there are many) is plausible, it remains unverified and experimentally unproven, as it is the ancient philosophical "ether" that fills the universe and through which light propagates. In this book and for all practical reasons, we adapt the clas-

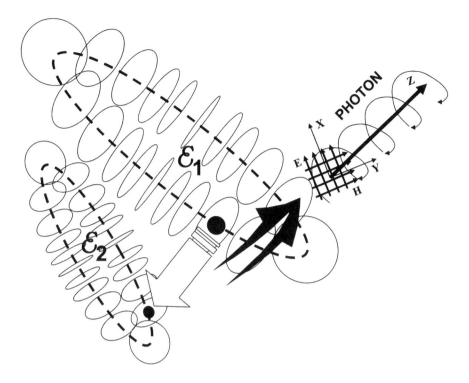

Figure 2.2. A photon is born. As the electron "transits" to a lower-energy orbit, the energy difference $E = E_1 - E_2 = h\nu$ is emitted in the form of a photon with a magnetic field (H) remaining orthogonal to electric field (E) and moving in a clockwise direction in a corkscrew-like manner.

sical theory of the dual nature of light, waves, and particles, as this has been proven in the lab of nature and explains most of the phenomena that involve light.

Based on the prevailing theory, if we could "see" what these unimaginable tiny "fuzzy" balls look like, we would discover that *all photons are not equal*. Some are red, some green, some blue, some have exotic colors, and some are even "invisible." Being different, photons have their own idiosyncratic moods and as they cross fields of forces within matter, each type of photon travels at its own pace and on its own individual path. And as they transverse matter, they also interact differently with it and among themselves. This has been managed and specific devices have been developed that are applicable in telecommunications and other applications.

In optical communications, a relatively good medium for photons to propagate through is glass, especially when it is in the form of a very thin, uniform, ultrapure fiber. In fact, glass fiber is the chosen transmission medium for high-speed, high-reliability, and long-distance communications, both terrestrial and submarine. With *Dense Wavelength Division Multiplexing* (DWDM) technology, 160 wavelengths have been commercially used in a single fiber, resulting to an aggregate bandwidth

30 OPTICAL PROPAGATION

Figure 2.3. An old and simple but convincing experiment: when exposed to light, pressure (due to absorption) is exerted on the blackened surface of the paddles and the wheel rotates according to action–reaction, proving that photons behave like particles.

that exceeds one terabit per second. DWDM systems are currently on a trend toward not only increasing both the density of wavelengths per fiber and thus the aggregate bit rate, but also the fiber span without signal amplification.

Certain numbers mentioned in this book may seem very small or very large. Table 2.1 puts in perspective what these number mean, such as a picosecond (ps) or a terabit per second (Tbps).

2.1.1 The Wave Nature of Light

Monochromatic light consists of waves all of exactly the same "single" frequency and it propagates in space with a spherical front. However, to simplify the mathematical wave analysis, electromagnetic waves are considered planar. Thus, light is described by the Maxwell's electromagnetic plane wave equations:

$$\nabla^2 \mathbf{E} = (1/v^2)\,(\partial^2 \mathbf{E}/\partial t^2)$$

$$\nabla^2 \mathbf{H} = (1/v^2)\,(\partial^2 \mathbf{H}/\partial t^2)$$

$$\nabla \mathbf{D} = \rho$$

$$\nabla \mathbf{B} = 0$$

Table 2.1

Number	10^N	Name	Time unit	Bit rate	Length unit
1,000,000,000,000,000	10^{15}	one quadrillion	(See note 1)	Peta-bps (Pbps)	
1,000,000,000,000	10^{12}	one trillion		Tera-bps (Tbps)	
1,000,000,000	10^9	one billion		Giga-bps (Gbps)	
1,000,000	10^6	one million		Mega-bps (Mbps)	
1,000	10^3	one thousand		Kilo-bps (Kbps) (See note 3)	Kilometer (km)
100	10^2	one hundred			
10	10^1	ten			Decameter (Dm)
1	10^0	one	Second (s) (See note 2)	Bit-per-second (bps)	Meter (m)
0.1	10^{-1}	one-tenth			Decimeter (dm)
0.01	10^{-2}	one-hundredth			Centimeter (cm)
0.001	10^{-3}	one-thousandth	Millisecond (ms)		Millimeter (mm)
0.000001	10^{-6}	one-millionth	Microsecond (μs)		Micrometer (μm)
0.000000001	10^{-9}	one-billionth	Nanosecond (ns)		Nanometer (nm) (See note 4)
0.000000000001	10^{-12}	one-trillionth	Picosecond (ps)		Picometer (pm)
0.000000000000001	10^{-15}	one-quadrillionth	Femtossecond (fs)		

1: Blank entries signify uncommon unit.
2: One day has 86,400 seconds. Light travels at about 10^{10} cm/second or 30 cm/ns.
3: The bit rate of uncompressed PCM voice is 64 Kbps.
4: 3.4 nm is the length of one helical turn of a DNA strand.

where ∇^2 is the Laplacian operator; v is the speed of the wave in a medium; θ is the partial derivative; **E** and **H** are the electric and magnetic fields, respectively; **D** is the electric displacement vector (its gradient is the charge density ρ); and **B** is the magnetic induction vector.

When an electromagnetic wave propagates in a linear medium (e.g., noncrystalline), these four vectors are inter-related by

$$\mathbf{D} = \varepsilon_0 \mathbf{E} + \mathbf{P}$$

$$\mathbf{B} = \mu_0 \mathbf{H} + \mathbf{M}$$

where ε_0 and μ_0 are the dielectric permitivity and permeability, respectively, both constants of free space, and **P** and **M** are the electric and magnetic polarization of the wave, respectively. Then, the electric polarization is expressed as

$$\mathbf{P} = \varepsilon_0 \chi \mathbf{E}$$

where χ is the electric susceptibility of the medium. In a nonlinear medium, this is expressed as a tensor and the latter relation includes higher-order terms. Moreover, the dielectric constant of the material, ε, is connected with the susceptibility as

$$\varepsilon = \varepsilon_0 (1 + \chi)$$

The propagation of a plane wave is described by the two relationships

$$\mathbf{E}(r, t) = \mathbf{E}_0 e^{-j(\omega t - \mathbf{k} \cdot \mathbf{r})}$$

and

$$\mathbf{H}(r, t) = \mathbf{H}_0 e^{-j(\omega t - \mathbf{k} \cdot \mathbf{r})}$$

where ω is the angular frequency, **r** is the directional vector, and **k** is the wave vector, which is connected with the dielectric constant by the relationship

$$k = |\mathbf{k}| = \omega \theta(\mu_0 \varepsilon)$$

The aforementioned two wave relationships are complex numbers and they consist of a real term, $\cos(x)$, and an imaginary term, $\sin(x)$. However, the real part of them is simplified to

$$\mathbf{E}(r, t) = \mathbf{E}_0 \cos(\omega t - \mathbf{k} \cdot \mathbf{r})$$

and

$$\mathbf{H}(r, t) = \mathbf{H}_0 \cos(\omega t - \mathbf{k} \cdot \mathbf{r})$$

When the (monochromatic) plane wave travels in free space, it travels at a maximum and constant speed (since μ_0 and ε_0 for free space are constant quantities):

$$c = \omega/k = 1/\sqrt{\mu_0 \varepsilon_0}$$

where $c = 2.99792458 \times 10^{10}$ cm/sec, or ~30 cm/nsec.

When light travels in a medium (other than free space), then its velocity u is expressed by

$$u = \omega/\mathbf{k} = 1/\sqrt{\mu_0 \varepsilon_0}$$

The speed of light, u, in a medium is always smaller than c, since $\mu > \mu_0$ and $\varepsilon < \varepsilon_0$.

2.2 THE PARTICLE NATURE OF LIGHT

Electromagnetic fields with electric field **E** and magnetic field **H** have an energy flux density F calculated as

$$F = (c/4\pi)[\mathbf{EH}]$$

The momentum of the field is calculated as

$$P = \int \{(1/4\pi c)[\mathbf{EH}]dV\}$$

Now, when this field is absorbed by a surface, the momentum transmitted normally to the unit of surface in the unit of time is the pressure exerted on it. Thus, light, like nuclear particles, exerts pressure and causes a wheel to spin (Figure 2.3). As a consequence, light is also described in terms of the number of particles or *photons*, the smallest quantity of monochromatic light, described by the energy (E) equation:

$$E = h\nu$$

where h is Planck's constant, $6.6260755 \times 10^{-34}$ (Joule-second), and ν is the frequency of light (hertz).

In free space, light (of all wavelengths) travels in a straight path at a constant maximum speed. However, the speed of light changes when it travels in a medium, and this change is not the same for all media or for all wavelengths. By free space is meant space that is completely free from matter (absolute vacuum) and from electromagnetic fields. Frequency, ν, speed of light in free space, c, and wavelength, λ, are interrelated by:

$$\nu = c/\lambda$$

2.3 CLASSICAL INTERFERENCE

When two monochromatic and coherent light sources (Figure 2.4) illuminate a screen, then, based on the Huygens–Fresnel principle, alternating bright and dark zones or fringes, are seen on the screen. The bright or dark state of a fringe depends

34 OPTICAL PROPAGATION

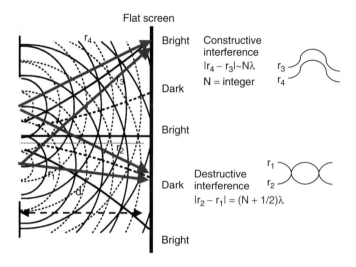

Figure 2.4. Two coherent monochromatic light sources create alternating bright and dark zones (fringes) on a screen, due to constructive and destructive interference, respectively.

on the absolute difference of the optical paths, the magnitude of which equals the travel distance:

$$\Delta = |\mathbf{r}_2 - \mathbf{r}_1| \text{ or } \Delta = |\mathbf{r}_4 - \mathbf{r}_3|$$

Bright zones (*constructive interference*) are formed when the path difference is an integer multiple of λ (i.e., $k\lambda$, $k = 1, 2, 3 \ldots$) and dark zones (*destructive interference*) are formed when the travel difference between the two rays is the half-integer multiple of λ [i.e., $(k + \frac{1}{2})\lambda$, $k = 0, 1, 2, 3 \ldots$].

2.4 QUANTUM INTERFERENCE

Consider an atom with only one electron at the outer orbit [e.g., Sodium (Na)] and on the lowest energy level or state S_1. Then, when a photon of energy E_{12} is absorbed, the electron "jumps" to a higher energy level or state S_2. While at level S_2, the electron may absorb another photon of energy E_{23} to "jump" to energy level or state S_3.

Photon absorption by the electron of an atom implies that the atom is "opaque" for that frequency (or wavelength) of photon. Photons that are not absorbed travel through it and, thus, the atom is "transparent."

Thus, when the atom is at level S_1, it is transparent to E_{23} and opaque to E_{12}, and when at level S_2, it is transparent to E_{12} and opaque to E_{23}. At some random time, and while at energy level E_3, the atom may transit to the lowest energy level S_1, releasing photonic energy.

Now, consider a "cloud" of such atoms illuminated with both E_{12} and E_{23}. Since all atoms in the cloud are neither excited by photons at the same time nor synchronized, jumping from one state to another becomes a random process. Statistically, there will be a time when a mix of both states will exist, S_1 and S_2. It turns out that, under a specific mix of states S_1 and S_2 and intensities E_{12} and E_{23}, the "cloud" seems to be transparent to both photonic energies (and, under different conditions, opaque). Macroscopically, this seems to be destructively interferometric but at the quantum (atomic energy) level; hence the term quantum interferometry. Quantum interferometry is an area of current research from which more findings and developments are expected in the near future.

2.5 LIGHT ATTRIBUTES

The attributes of light and their significance of interest in this book are shown in Table 2.2.

2.6 MATTER

The study of matter is extremely complex and almost impossible to be covered in a section of a book. Matter may be in one of several phases; gaseous, plasma (or completely ionized), liquid, and solid. For example, oxygen is typically a gas, but it is also liquid at low temperatures. Water (H_2O) is typically liquid, but at low temperatures it is solid and at high temperatures it is a gas. Moreover, each element and each compound exhibits different behavior and properties when in one of these phases. Some matter allows most optical energy (most photons) to propagate through it and it is called *optically transparent* (for example, glass or water). In

Table 2.2

Attribute	Significance
Dual nature	Electromagnetic wave and particle
Can have one of many λs	Wide and continuous spectrum
Polarized	Circular, elliptic, linear (TE_{nm}, TM_{nm}); affected by fields and matter
Optical power	Light can be in a wide power range (from μW to MW); affected by matter
Propagation	In free space it propagates in straight path; in matter it is affected differently (absorbed, scattered, passed through); in optical waveguides (fiber) it follows the bends
Propagation speed	In free space $c \sim 10^{10}$ cm/sec, in matter c/n; different wavelengths travel at different speeds
Phase	Affected by variations in fields and matter

contrast, some dense matter absorbs (or scatters) light within the first few atomic layers and it is called *nonoptically transparent* or *opaque* (for example, a sheet of metal).

Some matter passes a portion of optical energy through it and absorbs part of it (typically ~50%) and is called *semitransparent* (for example, most transparent matter, semitransparent mirrors). Such matter attenuates the optical power of light and may be used in optical devices known as *optical attenuators*.

Some matter allows selected frequencies to pass through and thus it is *optically transparent* to them; it absorbs all others. For example, red, green, yellow, or blue glass each allow a selected range of frequencies to be propagated through it; the Sun's ionized surface or hot sulfur vapors absorb specific frequencies. Such matter is called an *optical filter*.

Some matter permits rays with a certain polarization to propagate through it and absorbs or reflects the others (for example, polarizing sunglasses). Such matter is called a *polarizing filter*.

Some matter emits photons when in an intense electric field and some matter emits photons of specific wavelengths when it is illuminated with light of shorter wavelength (for example, most minerals under UV light; fluorescent substances, erbium, etc.).

Some matter is *homogeneously* optically transparent; that is, it has the same chemical, mechanical, electrical, magnetic, or crystallographic properties throughout its volume and in all directions.

Some matter is *heterogeneously* optically transparent; that is, it does not have the same chemical, mechanical, electrical, magnetic, or crystallographic properties throughout its volume.

Some matter is *isotropically* optically transparent; that is, it has the same index of refraction, polarization state, and propagation constant in every direction throughout the material.

Some matter is *anisotropically* optically transparent; that is, it does not have the same index of refraction, polarization state, and propagation constant in every direction throughout the material. Anisotropy is explained as follows: the electrons of certain crystals, such as calcite ($CaCO_3$), move with different amounts of freedom in selective directions in the crystal, and the dielectric constant as well as the refractive index of the crystal is different in these selective directions. As a result, as photons enter the crystal their electromagnetic fields interact differently in one direction than in another, and this affects their propagation pattern in the crystal.

2.6.1 Attributes of Matter

The attributes of matter and their significance of interest to optical communications DWDM may be summarized as follows:

- Refractive index (n)   Is a function of molecular structure of matter
  Is a function of optical frequency $\{n(\omega)\}$
  Is a function of optical intensity

	Determines optical propagation properties of each λ
	May not be distributed equally in all directions
	Is affected by external temperature, pressure, and fields
• Reflectivity (*R*)	Is a function of geometry, λ, and *n*
	Material surface reflects optical power
	Changes polarization of incident optical wave
	Changes phase of incident optical wave
• Transparency (*T*)	Depends on matter consistency
• Scattering	Due to molecular matrix disorders
• Absorption (*A*)	Due to the presence of ions in the molecular matrix and atomic excitation; ions act like dipoles, which exhibit eigenfrequencies and antenna characteristics (receiver/transmitter)
• Polarization (*P*)	Due to X–Y uneven electromagnetic fields (light–matter interaction)
• Birefringence (*B*)	Due to nonuniform distribution of *n* in all directions
• Phase shift (ΔΦ)	Due to wave propagation property of light through matter

2.7 PROPAGATION OF LIGHT

As already described, when light enters matter, it is reflected by its surface and refracted by the matter, and its velocity changes as well as its wavelength, but not its frequency.

2.7.1 Reflection and Refraction—Snell's Law

Index of refraction of a transparent medium (n_{med}) is defined as the ratio of the speed of monochromatic light in free space, c, over the speed of the same monochromatic light in a medium (v_{med}):

$$n_{med} = c/v_{med}$$

Then, between two mediums (1 and 2) the following relationships are true:

$$n_2/n_1 = v_1/v_2$$

and

$$n_1 \cos \beta = n_2 \cos \alpha$$

where n_1, v_1, and n_2, v_2 are the index of refraction and speed of light in the two media, and α and β are the angle of incidence and angle of refraction, respectively.

When the angle of incidence is very small, $\cos \alpha = 1 - \alpha^2/2$ and the cosine equation becomes

$$n_1(1 - \beta^2/2) = n_2(1 - \alpha^2/2)$$

The index of refraction, or *refractive index,* for free space has the numerical value of 1, whereas for other materials it is typically between 1 and 2, and in some cases greater than 2 or 3.

The reflected portion of monochromatic light is known as Fresnel reflection. The amount of reflected power as well as the polarization state of the reflected light depends on the polarization state of the incident light, on the angle of incidence, and on the refractive index difference.

For normal incidence on a single surface the reflectivity, ρ, is given by the Fresnel equation:

$$\rho = (n - 1)^2/(n + 1)^2$$

If the absorption of the material over a length d is A, which is calculated from the absorption coefficient (absorbed power per centimeter) α, then the internal material transmittance, τ_i, is defined as the inverse of the material absorption. For internal transmittance τ_i, the external input–output transmittance (taking into account the reflectivity, ρ, at the surface) is given by

$$\tau = [(1 - \rho)^2 \tau_1]/[1 - \rho^2 \tau_1^2]$$

The following basic relationships are also useful:

Speed of light in free space: $c = \lambda f$
Speed of light in medium: $v_{med} = \lambda_{med} f$
Index of refraction: $n_1/n_2 = \lambda_2/\lambda_1$

where f is the frequency of light. Both letters f and ν are used for frequency. Here, we use f to eliminate confusion between v (for speed) and ν (for frequency).

Snell's law relates the ratio of the index of refraction with the angle of the incident (Θ_i) and refracted (Θ_t) rays:

$$n_2/n_1 = \sin \Theta_i / \sin \Theta_t$$

The *critical angle,* $\Theta_{critical}$, is the (maximum) angle of incidence of light (from a material with high to low refractive index) at which light stops being refracted and is totally reflected. As the angle of incidence approaches the critical angle, the refracted ray becomes parallel to the surface (without added phase shift) and is said to be *evanescent.* Beyond that point, there is no refracted ray. The critical angle depends on the refractive index and the wavelength of light:

$$\sin \Theta_{critical} = n_1/n_2$$

for $n_1 = 1$ (air), then,

$$\sin \Theta_{critical} = 1/n_2$$

In certain cases, a gradual variation of refractive index may take place. When light rays enter from one side, then rays are refracted such that they may emerge from the same side from which they entered. This is the case of the natural phenomenon known as a *mirage*.

2.7.2 Phase and Group Velocity

A monochromatic (single ω or λ) wave that travels along the fiber axis is described by

$$E(t, x) = A \exp[j(\omega t - \beta x)]$$

where A is the amplitude of the field, $\omega = 2\pi f$, and β is the propagation constant.

Phase velocity, v_ϕ, is defined as the velocity of an observer that maintains constant phase with the traveling field, that is, $\omega t - \beta x =$ constant.

Replacing the traveled distance x within time t, $x = v_\phi t$, then the phase velocity of the monochromatic light in the medium is

$$v_\phi = \omega/\beta$$

When a signal is transmitted in a dispersive medium, it is necessary to know its speed of propagation. A continuous sine wave does not provide any meaningful information, as a real optical signal consists of a band of frequencies in a narrow spectrum. Moreover, each frequency component in the band travels (in the medium) with slightly different phase. This is explained mathematically as follows.

Consider an amplitude-modulated optical signal traveling along a fiber:

$$e_{AM}(t) = E[1 + m \cos(\omega_1 t)]\cos(\omega_c t)$$

Where E is the electric field, m is the modulation depth, ω_1 is the modulation frequency, ω_c is the frequency of light (or carrier frequency), and $\omega_1 \ll \omega_c$.

Trigonometric expansion of the above expression results in three frequency components with arguments:

$$\omega_c, \quad \omega_c - \omega_1, \quad \text{and} \quad \omega_c + \omega_1$$

Each component travels along the fiber at slightly different phase velocity (β_c, $\beta_c - \Delta\beta$, $\beta_c + \Delta\beta$, respectively), accruing a different phase shift. Eventually, all three components form a spreading envelope that travels along the fiber with a phase velocity:

$$\beta(\omega) = \beta_c + (\partial\beta/\partial\omega|\omega = \omega_c)\Delta\omega = \beta_c + \beta'\Delta\omega$$

where ∂ denotes a partial derivative.

Group velocity, $v_g = c/n_g$, considers a group of frequencies within a band, $\omega_0 - \Delta\omega/2 < \omega < \omega_0 + \Delta\omega/2$. The frequency band $\Delta\omega$ is considerably smaller than the carrier frequency ω_0. The group velocity is defined as the velocity of an observer that maintains constant phase with the group traveling envelope. Mathematically, the group velocity is defined as the derivative of the frequency ω with respect to the propagation constant β:

$$v_g = \partial\omega/\partial\beta = 1/\beta'$$

where β' is the first partial derivative with respect to ω (remember that in this case ω in the frequency band is not purely monochromatic). Notice that the group velocity is a function of frequency; this will be useful in the discussion on dispersion.

2.7.3 Spectral Broadening

The refractive index depends on the electrical field. Thus, as an almost monochromatic light pulse travels in a transparent medium, its amplitude variation causes *phase change* and *spectral broadening*. Notice that the group velocity is a function of frequency; this will be useful in the discussion on dispersion.

Spectral broadening causes one-half of the pulse to be frequency downshifted (known as *red shift*) and the other half to be frequency upshifted (known as *blue shift*). Such shifts are also expected in pulses that consist of a narrow range of wavelengths that are centered on the zero-dispersion wavelength. Below the zero-dispersion point, wavelength dispersion is negative and above it wavelength dispersion is positive.

2.7.4 Scattering and Absorption

In the process of manufacturing optical fiber and during the purification process, certain undesired elements are not completely removed. These elements, in addition to lattice defects, alter the optical characteristics of transparent matter, causing absorption and/or scatter of photons that results in optical throughput loss. A similar effect is also caused by small fluctuations of the refractive index when they are on the order of the wavelength of light passing through it. This particular effect is known as *Rayleigh scattering*.

2.7.5 Microcracks

Microcracks in the crystallized matrix of matter or in amorphous solid matter are generated due to stresses (mechanical or thermal) or due to material aging. They may have an adverse effect on the propagation of light as well as on the strength of the material. In fact, if the crack gap (the distance between the crack surfaces) increases due to tensile forces on the order of a wavelength ($\lambda/4$, $\lambda/2$) it may act as an uninvited filter.

2.7.6 Mechanical Pressure

Mechanical pressure applied on matter disturbs its internal microstructure and causes local variation of the refractive index. This variation affects the propagation of light in matter. Similarly, when fiber is bent, the outer periphery experiences stretching and the inner periphery experiences compression, and this also affects optical propagation. Thus, the safe bend radius* recommended by ITU-T is 37.5 mm (ITU-T G.652, paragraph 5.5, note 2).

2.7.7 Temperature Variation

Temperature variations affect the physical, mechanical, electrical, magnetic, chemical, and crystalline (if any) properties of matter. As a result, they affect the dielectric constant and the refractive index, which affect the propagation of light. Similar effects have temperature gradients along the length of the fiber.

2.8 DIFFRACTION

Consider a screen having a small hole with sharp edges at the periphery and dimensions comparable to the wavelength of light (this may be an aperture, a slit, or a grating) and a second screen behind it at distance d (Figure 1.18). Consider a parallel (collimated) and monochromatic beam, known as *light from a source at infinity*, directed to the small hole. Since light travels in free space in a straight path, a small round projection is expected, D_{EXP}. However, a wider projection is seen instead, D_{ACT}, with concentric rings surrounding it. The smaller the hole diameter, the wider the projection.

This phenomenon is attributed to interference at each edge point of the hole, which acts like a secondary source resulting in *diffraction;* this is also known as the *phenomenon of Fresnel*. Moreover, the intensity of rings, J, decreases with distance from the center of the projection (Figure 2.5). This pattern is also known as *Fraunhofer diffraction*.

According to Huygens's principle, the incident wave excites coherent secondary waves at each point of the wavefront. All these waves interfere with each other and cause the diffraction pattern. The pure mathematical analysis of diffraction is studied by approximating the scalar Helmholz equation in which the wavefront is a spherical function. However, to illustrate this point, if a plane wave is incident on

---

*This may also vary depending on whether the cable is specified to be under tension or not. Manufacturers usually specify a safe bend radius in the data sheets of their cables.

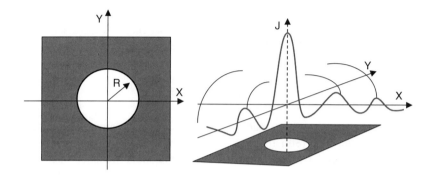

Figure 2.5. Intensity distribution of diffracted light passing through a round hole.

an aperture, then the radiant intensity per unit of projected area of the source viewed from that direction (that is the radiance, L), is described by

$$L(\alpha, \beta - \beta_0) = \gamma_0(\lambda^2/A)[F\{U(x, y, z = 0)\exp(i2\pi\beta_0 y)\}]^2 \quad (W/\text{sr-m}^2)$$

where α, β, and γ are propagation vectors in the x, y, and z directions, respectively, A is the area of the diffracting aperture, λ is the wavelength, F is the Fourier transform, and U is the complex amplitude distribution emerging from the aperture (hence, $z = 0$).

Integrating the above, invoking the conservation of energy theorem, and using Parceval's (or Rayleigh's) theorem, which converts the integral of the squared modulus of a function $f(x)$ to the integral of the squared modulus of its Fourier transform $f(\omega)$,

$$\int |f(x)|^2 dx = (1/2\pi)\int |f(\omega)|^2 d\omega$$

where the integrals (int) are from $-$ to $+$ infinity, the radiance is expressed by

$$L(\alpha, \beta - \beta_0) = K\gamma_0(\lambda^2/A)|F\{U(x, y, z = 0)\exp(i2\pi\beta_0 y)\}|^2 \quad \text{for } a^2 + b^2 \leq 1$$

$$L(\alpha, \beta - \beta_0) = 0 \quad \text{for } a^2 + b^2 > 1$$

and K is a normalization constant.

Now, let us consider an aperture that is a narrow rectangular slit with height h and width w. In this case, the monochromatic collimated beam passes through it and, due to refraction, the projection on a screen is a rectangle rotated by 90°. That is, the refracted pattern is narrow in the direction in which the aperture is wide. Moreover, because of two-dimensional Fourier expansion, the refracted light forms many secondary rectangles in the X–Y plane of the screen, with intensity fading away from the axis of symmetry (Figure 2.6). This is also known as *Fraunhofer diffraction* of a rectangular aperture. The condition for these rectangles on the screen is

$$R(x, y) = \text{Rect}(x/w_0)\text{Rect}(y/h_0) = 1 \quad \text{for } |x| < w_0/2 \text{ and } |y| < h_0/2$$

and

$$R(x, y) = 0 \quad \text{elsewhere}$$

Theoretically speaking, when a diffracting aperture is illuminated by a uniform monochromatic plane wave, λ, the total radiant power, P_{TR}, emanating from the diffracting aperture is calculated using Rayleigh's theorem and Fourier transforms:

$$P_{TR} = \lambda^2 \int_{-\infty}^{\infty} \left[\int_{-\infty}^{\infty} |U_0(x, y; 0)e^{i2\pi\beta y}|^2 dx \right] dy$$

where U_0 is the function of the complex amplitude distribution emerging from the diffracting aperture, β is the propagating vector and $\{x, y\}$ are the Cartesian coordinates.

The above discussion applies to monochromatic light. If light is polychromatic, then each color (frequency) component would be diffracted differently; that is, at a different angle. Then, fringes on the screen would be of different colors but overlapping. The corollary is that the angle of diffraction depends on frequency (color) of the light.

If in the above experiment, instead of one slit there is an array of parallel slits then, because of interference among the slits, light is diffracted at certain angles. This diffraction of light is known as *diffraction by transmission*. Interesting phenomena occur when slits are positioned in a matrix configuration. Diffraction then takes place in a two-dimensional plane and the fringe pattern depends on size, shape, density, wavelength, and topological arrangement of slits. Thus, the accuracy of diffraction-based devices may impact the spectral content of an optical signal, which may be manifested as signal-to-noise ratio and BER degradation.

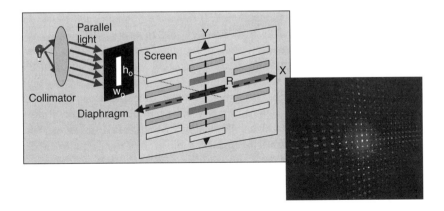

Figure 2.6. Diffraction of light passing through a rectangular hole (diagram and photo).

Now, if in the above discussion, instead of slits in the diaphragm there are parallel grooves each at a width comparable to the wavelength, then the same theory applies. The difference is that light diffracts back at specified angles. This is known as *diffraction by reflection*.

2.9 POLARIZATION

If we examine the electrical state of matter on a microscopic level, we discover that it consists of charges, the distribution of which depends on the presence or absence of external fields. If we assume that for every positive charge there is a negative charge, then we may think that each positive–negative charge constitutes an electric dipole. The electric moment of a dipole at some distance is a function of distance and charge density. Now, for a distribution of electric dipoles, the *electric dipole moment per unit volume* is defined as the *polarization vector* **P**.

Two relations describe the propagation of light in nonconducting media:

$$E(r, t) = \varepsilon_1 E_0 e^{-j(\omega t - \mathbf{k} \cdot \mathbf{r})}$$

and

$$H(r, t) = \varepsilon_2 H_0 e^{-j(\omega t - \mathbf{k} \cdot \mathbf{r})}$$

where ε_1 and ε_2 are two constant unit vectors that define the direction of each field, **k** is the unit vector in the direction of propagation, and E_0 and H_0 are complex amplitudes, which are constant in space and time.

Assuming a wave propagating in a medium without charges, then $\Delta E = 0$ and $\Delta H = 0$. Based on this, the product of unit vectors is

$$\varepsilon_1 \cdot \mathbf{k} = 0 \quad \text{and} \quad \varepsilon_2 \cdot \mathbf{k} = 0$$

That is, the electric (E) and the magnetic (H) fields are perpendicular to the direction of propagation **k** (Figure 2.2). Such a wave is called a *transverse wave*.

Polarization of electromagnetic waves is a complex subject, particularly when light propagates in a medium with different refractive indexes in different directions. As light propagates through a medium, it enters the fields of nearby dipoles and field interaction takes place. This interaction may affect the strength of the electric and/or magnetic fields of light differently in certain directions so that the end result may be a complex field with an elliptical or a linear field distribution. For example, the electric field *e* becomes the linear combination of two complex fields E_{0x} and E_{0y}, the two componets in the *x* and *y* directions of a Cartesian coordinate system, such that

$$E(r, t) = (\varepsilon_x E_{0x} + \varepsilon_y E_{0y}) e^{-j(\omega t - \mathbf{k} \cdot \mathbf{r})}$$

2.9 POLARIZATION

This relationship implies that the two components, E_{0x} and E_{0y}, vary sinusoidally, they are perpendicular to each other, and there may be a phase betwen them, ϕ. From this relationship, a vector is defined known as the *Jones vector;* the Jones vector $\mathbf{J} = [J_1, J_2]$ is related to the radiation aspects of the wave:

$$\mathbf{J} = \begin{vmatrix} E_{0x}\, e^{j\phi_x} \\ E_{0y}\, e^{j\phi_y} \end{vmatrix}$$

In this case, the dielectric quantity ε is described by a tensor that, in general, has different values in the three axes:

$$\varepsilon = \begin{vmatrix} \varepsilon_x & 0 & 0 \\ 0 & \varepsilon_y & 0 \\ 0 & 0 & \varepsilon_z \end{vmatrix} = \varepsilon_0 \begin{vmatrix} n_x^2 & 0 & 0 \\ 0 & n_y^2 & 0 \\ 0 & 0 & n_z^2 \end{vmatrix}$$

Now, from

$$\Delta^2 E = (1/v^2)(\partial^2 E/\partial t^2)$$

and

$$E(r, t) = \varepsilon E_0 e^{-j(\omega t - \mathbf{k} \cdot \mathbf{r})}$$

one obtains

$$\mathbf{k} \times (\mathbf{k} \times E_0) + \mu_0 \varepsilon \omega^2 E_0 = 0$$

or

$$[\mathbf{k} \times (\mathbf{k} \times I) + \mu_0 \varepsilon \omega^2][E_0] = 0$$

where *I* is the identity matrix. The latter is a vector equation equivalent to a set of three homogeneous linear equations with unknowns the components of E_0, E_{0x}, E_{0y}, and E_{0z}. In a typical case, the component E_{0z} along the axis of propagation is equal to zero.

This vector equation determines a relationship between the vector $\mathbf{k}$ (k_x, k_y, k_z), the angular frequency ω, and the dielectric constant ε (ε_x, ε_y, ε_z), as well as the polarization state of the plane wave.

Now, the term $[\mathbf{k} \times (\mathbf{k} \times I) + \mu_0 \varepsilon \omega^2]$ describes a three-dimensional surface. As the (complex) electric field is separated into its constituent components, each component may propagate in the medium at a different phase. The phase relationship as well as the magnitude of each vector defines the *mode of polarization.*

If E_{0x} and E_{0y} have the same magnitude and are in phase, then the wave is called *linearly polarized.*

If E_{0x} and E_{0y} have a phase difference (other than 90°), then the wave is called *elliptically polarized.*

If E_{0x} and E_{0y} have the same magnitude but differ in phase by 90°, then the wave is called *circularly polarized.* For example, in circularly polarized light the wave equation (propagating in the z direction) becomes

$$E(r, t) = E_0(\varepsilon_x \pm j\varepsilon_y)e^{-j(\omega t - \mathbf{k} \cdot \mathbf{r})}$$

Then, the two real components (in the x and in the y directions) are

$$E_x(r, t) = E_0 \cos(\mathbf{k} \cdot \mathbf{r} - \omega t)$$

and

$$E_y(r, t) = \pm E_0 \cos(\mathbf{k} \cdot \mathbf{r} - \omega t)$$

These equations indicate that at a fixed point in space the fields are such that the electric vector is constant in magnitude but it rotates in a circular motion at a frequency ω. The term $\varepsilon_x + j\varepsilon_y$ indicates a counterclockwise rotation (when facing the oncoming wave), and this wave is called *left-circularly polarized* or a wave with *positive helicity.* The term $\varepsilon_x - j\varepsilon_y$ indicates a clockwise rotation (when facing the oncoming wave), and this wave is called *right-circularly polarized* or a wave with *negative helicity.*

Using the notion of positive and negative helicity, E can be rewritten as

$$E(r, t) = (\varepsilon_+ E_+ + \varepsilon_- E_-)e^{-j(\omega t - \mathbf{k} \cdot \mathbf{r})}$$

where E_+ and E_- are complex amplitudes denoting the direction of rotation.

Now, if E_+ and E_- are in phase but have different amplitude, the last relationship represents an *elliptically polarized* wave with principal axes of the ellipse in the directions ε_x and ε_y.

Then, the ratio of the semimajor to semiminor axis is $(1 + r)/(1 - r)$, where $E_-/E_+ = r$.

If the amplitudes E_+ and E_- have a difference between them, $E_-/E_+ = re^{j\alpha}$, then the ellipse traced out by the vector **E** has its axes rotated by an angle $\phi/2$.

When $E_-/E_+ = r = \pm 1$, then the wave is *linearly polarized.* Thus, we have come to the same definition of polarization modes. The above discussion for the electric field E can also be repeated for the magnetic field **H**.

It can be shown that the displacement vector yields a vector equation similar to the one above as well as a refractive index with an ellipsoidal distribution or an ellipsoid of revolution around the z axis (the axis of propagation). The ellipsoidal distribution is the result of the displacement vector **D** and the electric field E, which are related by

$$\mathbf{D} = \varepsilon_0 \varepsilon E$$

where ε is a tensor (a 3 × 3 matrix with elements ε_{xx}, ε_{xy}, ε_{xz}, ε_{yx}, ε_{yy}, ε_{yz}, ε_{zx}, ε_{zy}, ε_{zz}). Because of the conservation of energy, the tensor is symmetric, that is $\varepsilon_{xy} = \varepsilon_{yx}$, $\varepsilon_{xz} = \varepsilon_{zx}$, and $\varepsilon_{yz} = \varepsilon_{zy}$, the mathematical derivation of the tensor expression in the three displacement components is greatly simplified:

$$D_x = \varepsilon_0 \varepsilon_x E_x, \quad D_y = \varepsilon_0 \varepsilon_y E_y, \quad \text{and} \quad D_z = \varepsilon_0 \varepsilon_z E_z$$

Based on this, the energy, W, stored in the electric field is then expressed by

$$W = \tfrac{1}{2} DE = \varepsilon_0 (D_x^2/\varepsilon_x + D_y^2/\varepsilon_y + D_z^2/\varepsilon_z)$$

In this relationship, the parenthetical term is of the form $(x^2/a + y^2/b + z^2/c)$, which defines an ellipsoid, and, hence, the name of the ellipsoidal refractive index of nonlinear optical crystals. Now, remember that a cross section through the center of an ellipsoid can yield an ellipse or a circle. In addition, a sphere for which $a = b = c$ is a special case of an ellipsoid (this is, also the definition of isotropy).

In simple terms, this can be summarized as follows. The electromagnetic wave nature of monochromatic light implies that the electric and the magnetic fields are in quadrature and in time phase. When created light propagates in free space, the two fields change sinusoidally and are perpendicular to each other. When light enters matter, then, depending on the displacement vector distribution in matter (and hence the dielectric and the refractive index), light (its electric and/or magnetic field) interacts with it in different ways. If the two planes (of the electric or magnetic field) are fixed in a Cartesian coordinate system, then light is *linearly polarized*. If on the other hand, the planes keep changing in a circular (or helicoidal) motion and the fields remain at the same intensity, then light is *circularly polarized*. If the intensity of the field changes monotonically, then light is *elliptically polarized*. Now, if we consider that light is separated into two components, one linearly polarized, I_P, and one unpolarized, I_u, then, the degree of polarization, P, is defined as

$$P = I_P/(I_P + I_U)$$

Light may be polarized when it is reflected, refracted, or scattered. In polarization by reflection, the degree of polarization depends on the angle of incidence and on the refractive index of the material, given by *Brewster's law:*

$$\tan(I_P) = n$$

where n is the refractive index and I_P the polarizing angle.

2.9.1 The Stokes Vector

The Jones vector describes the polarized radiation components of the electromagnetic wave. However, in the general case, the propagating wave may consist of a polarized and a nonpolarized component. This is described by the Stokes vector, **S** = [S_0, S_1, S_2, S_3], in which each of the four parameters are defined in terms of the two component fields E_{0x} and E_{0y} and the phase difference, ϕ, as

$$S_0 = E_{0x}^2 + E_{0y}^2$$

$$S_1 = E_{0x}^2 - E_{0y}^2$$

$$S_2 = 2E_{0x}E_{0y} \cos \phi$$

$$S_3 = 2E_{0x}E_{0y} \sin \phi$$

and are related as

$$S_0^2 = S_1^2 + S_2^2 + S_3^2$$

Based on this definition, light intensity, degree of polarization, ellipticity, amplitude ratio of the field components in the x and y directions and their phase difference, and polarization azimuth are expressed in terms of the Stokes parameters as

$$I \text{ (Intensity)} = S_0$$

$$P \text{ (degree of polarization)} = \sqrt{(S_1^2 + S_2^2 + S_3^2)}/S_0$$

$$\varepsilon \text{ (ellipticity)} = \tfrac{1}{2} \arctan [S_3/\sqrt{S_1^2 + S_2^2}]$$

$$E_{0y}/E_{0x} \text{ (amplitude ratio)} = \sqrt{(S_0 - S_1)/(S_0 + S_1)}$$

$$\phi \text{ (phase difference)} = \phi_y - \phi_x = \arccos [S_2/(S_0^2 - S_1^2)]$$

$$\alpha \text{ (azimuth)} = \tfrac{1}{2} \arctan (S_2/S_1)$$

Relationship between Jones and Stokes Vectors. The relationship for specific polarization states of the Jones and Stokes vectors are:

- Unpolarized light: **J** = [not defined], **S** = [1, 0, 0, 0]
- Circular polarized right: **J** = $\sqrt{2^{-1}}$[1, i], **S** = [1, 0, 0, 1]
- Circular polarized left: **J** = $\sqrt{2^{-1}}$[1, $-i$], **S** = [1, 0, 0, -1]
- Linear polarization horizontal: **J** = [1, 0], **S** = [1, 1, 0, 0]
- Linear polarization vertical: **J** = [0, 1], **S** = [1, -1, 0, 0]
- Linear polarization +45°: **J** = $\sqrt{2^{-1}}$[1, 1], **S** = [1, 0, 1, 0]
- Linear polarization $-45°$: **J** = $\sqrt{2^{-1}}$[1, -1], **S** = [1, 0, -1, 0]

2.9.2 The Poincaré Sphere

The interaction of light with matter affects the state (or states) of polarization (SoP). In general, a SoP with azimuth α and ellipticity ε is expressed by

$$\text{SoP} = \begin{vmatrix} 1 + \cos(2\alpha)\cos(2\varepsilon) \\ \cos(2\alpha)\sin(2\varepsilon) + i\sin(2\alpha) \end{vmatrix}$$

Under certain circumstances, the SoP of a wave changes (rotates) as it travels through a medium; such change is associated with delay. When the rotation is 90°, it results in two orthogonal SoPs (one before and one after the rotation). Two mutually orthogonal SoPs, both at equal intensity, result in a depolarized field.

Described in terms of the Stokes parameters, the unpolarized part $S_0 - \sqrt{S_1^2 + S_2^2 + S_3^2}$ in the Poincaré sphere is left out and only the polarized part is considered. The three Stokes parameters on terms of the azimuth χ and ellipticity ψ are expressed by

$$S_1 = \cos(2\varepsilon)\cos(2a)$$

$$S_2 = \cos(2\varepsilon)\sin(2a)$$

$$S_3 = \sin(2\varepsilon)$$

The Poincaré sphere is a graphical representation of all possible SoPs; each point on the sphere's surface is a SoP defined by **S** (Figure 2.7).

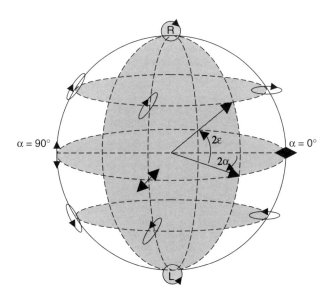

Figure 2.7. Poincaré sphere.

2.9.3 Birefringence

Anisotropic materials have a different index of refraction in specific directions. As such, when a beam of monochromatic unpolarized light enters the material at a specific angle and travels through it, it is refracted differently in the directions of different indices (Figure 2.8). That is, when an unpolarized ray enters the material and is separated into two rays, each with different polarization, different direction, and different propagation constant; one is called the *ordinary* ray (O) and the other the *extraordinary* ray (E). In these directions the refracted index is similarly called the *ordinary index,* n_0, and the *extraordinary index,* n_e, respectively. This property of crystals is known as *birefringence.* In general, all optically transparent crystals have some degree of birefringence (some more than others), unless they belong to the cubic system or they are amorphous.

Certain optically transparent isotropic materials become anisotropic when under stress. Stress may be exerted due to mechanical forces (pulling, bending, twisting), due to thermal forces (ambient temperature variations), and due to electrical fields. Under such conditions, the index of refraction and polarization and propagation characteristics become different in certain directions within the material.

Clearly, birefringence in fiber alters the polarization state of the characteristics of the propagating optical signal, and several techniques have been developed to address this phenomenon such as *polarization spreading* (polarization scrambling, data-induced polarization), or *polarization diversity.*

2.9.4 Polarization-Dependent Loss

When light travels through matter it suffers power loss. One of the contributors to power loss is polarization. Virtually all optically transparent materials affect to

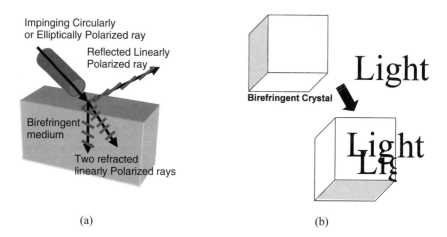

(a) (b)

Figure 2.8. (a) Birefringence in a nonpolarized beam. (b) An example of what happens when a birefringent crystal is placed on top of the word "Light."

some degree the polarization state of light. In general, optically transparent materials exhibit a spatial polarization distribution. As an optical signal passes through it, it suffers power reduction or optical power loss in selective directions due to spatial polarization interaction; this loss is wavelength dependent and is known as *polarization-dependent loss* (PDL). It is measured in decibels (dB). PDL affects the signal quality and system performance.

At low data rates, PDL is a minor contributor to loss. However, at 10 Gbit/s and above, PDL becomes comparable to insertion loss (IL). Therefore, at high bit rates PDL needs careful examination. The surprising result here is that over a span with many connectors the total loss due to PDL is not an algebraic sum. To explain this, consider two cascaded PDL elements, A and B. Element A attenuates the optical signal due to a partial polarization rotation (PDL_A). This partially distorted signal enters element B, which has a random polarization orientation and properties with element A. Therefore, the partially polarized light from element A undergoes another stage of polarization distortion and loss as it enters B (PDL_B). As an example, if the two elements have parallel orientations and similar characteristics, the total PDL is $L_{PDL,1-2} = (L_{PDL,1} + L_{PDL,2})/(1 + L_{PDL,1} L_{PDL,2})$.

The power loss due to polarization of the received signal may vary by ~0.5 dB. In worst-case optical transmission design, the maximum loss level (~ –0.5 dB) should be used, whereas in typical performance design an average level (~ –0.1 dB) is used. This value does not change with respect to the center wavelength of the received signal. However, asymmetric spectral polarization loss causes asymmetric amplitude signal distortions and the signal may appear with shifted center wavelength.

2.9.5 Extinction Ratio

Consider polarized light traveling through a polarizer. The maximum transmittance, T_1, is termed the *major principal transmittance,* and the minimum, T_2, is termed the *minor principal transmittance.* The ratio of major to minor is known as the *principal transmittance.* The inverse, minimum to maximum, is known as the *extinction ratio.*

Consider two polarizers in tandem, one behind the other, with parallel surfaces. Then, if their polarization axes are parallel, the transmittance is $T_1^2/2$. If their axes are crossed (perpendicular), the transmittance is $2T_2/T_1$. This is also (but erroneously) termed the *extinction ratio.*

In optical communications, the term extinction ratio is defined slightly differently. It describes the modulation efficiency in the optical medium, considering either that the laser source is directly modulated or that it is always continuous and externally modulated on–off (i.e., on–off keying modulation). In this case, the extinction ratio is defined as the ratio of the transmitted optical power of a logic 1 (on), P_1, over the transmitted optical power of a logic 0, P_0, and it is measured in decibels:

$$\text{Extinction ratio} = 10 \times \log(P_1/P_0) \text{ (dB)}$$

or, in percentage:

$$\text{Extinction ratio} = (P_1/P_0) \times 100 \text{ (\%)}$$

2.9.6 Phase Shift

Dielectric materials, in contrast to polarization rotation, shift the phase of light transmitted through them. The amount of phase shift $\Delta\phi$ depends on the wavelength λ, the dielectric constant ε, the refractive index ratio n_1/n_2, and the optical path (thickness) of material.

2.10 PARADOXES

2.10.1 Background

In this section, we make certain observations that appear to be paradoxes, which, if proven feasible, may open a new chapter in optical and photonic technology and optical propagation.

In the most general solution, the electric field of a monochromatic traveling plane harmonic wave in free space is of the form $E(z, t)$:

$$E(z, t) = Re\{\mathbf{F} \exp[-i\omega(t - \mathbf{rn}/c)]\}$$

Where Re is the real part of $\{\ \}$, $\mathbf{F}$ is a complex vector (of the form $\alpha + i\beta$), ω is the wave frequency, $\mathbf{n}$ is a unit vector along the axis of propagation z, such as $z = \mathbf{rn}$, and $\mathbf{k} = \omega\mathbf{n}/c$ is the wave vector such that $\mathbf{k} = 2\pi\mathbf{n}/\lambda$, $\lambda = 2\pi/\mathbf{k}$, and $\lambda = 2\pi c/\omega$.

When light propagates in matter, then

$$dH/dz = -i(\omega/c)(\varepsilon E), \text{ and } dE/dz = -i(\omega/c)(\chi H)$$

where ε and χ are the dielectric constant and susceptibility of the material.

From the last relationship we obtained

$$d^2E/dz^2 = -(w^2/c^2)(\varepsilon\chi E)$$

The solution of this is

$$E(z, t) = Re\{F \exp[i(\omega/c)\sqrt{\varepsilon\chi}z - i\omega t)]\}$$

In this case, the wave vector is $\mathbf{k} = (\omega/c)\sqrt{\varepsilon\chi}z$, or $\mathbf{k} = (\omega/v)z$, where v is the speed of light in the medium.

In this analysis, we consider monochromatic light. Realistically, the frequency of a photonic signal consists of a band of frequencies distributed within $\Delta\omega$, $\omega_0 - \Delta\omega/2 < \omega < \omega_0 + \Delta\omega/2$. Introducing a new variable, $\xi = \omega - \omega_0$, this yields a term $E_0 \exp[-i(\omega_0 t - \mathbf{k}_0 z)]$ that represents a traveling wave with a mean carrier frequency ω_0 but with an amplitude that is no longer constant but has a maximum value propagating with group velocity $u_g = d\omega/d\mathbf{k}$.

2.10.2 Paradoxes

One consequence of this analysis is that the spectral width $\Delta\omega$ of the signal and the duration of the signal (or period) Δt set a minimum value:

$$\Delta\omega \cdot \Delta t|_{min} \sim 2\pi$$

The latter provides an estimation of the minimum spectral content of a signal as the data rate increases, a limit that is useful in photonic transmission engineering. A similar consequence is

$$\Delta\mathbf{k} \cdot \Delta x \sim 2\pi$$

Thus, a disturbance within a region Δx is related to a superposition of monochromatic waves that lie within a spectral range $\Delta\mathbf{k}$ of the order $2\pi/\Delta x$.

Another consequence is that the product of the phase velocity $u = \omega/\mathbf{k}$ and group velocity u_g is $uu_g = c^2$. Since the group velocity in a medium is always $u_g < c$, it turns out that $u > c$, a result that is contradicts many theories.

Another consequence is that the dielectric constant of the medium $\varepsilon(\omega)$ can be expressed in terms of the charge density Ne, the electron mass, m, and the difference of frequency squares $\omega_0^2 - \omega^2$, where ω_0 is the oscillation of charges (assuming elastic forces between them) and ω is the frequency of an oscillating field:

$$\varepsilon(\omega) = 1 + (4\pi Ne^2)/m(\omega_0^2 - \omega^2)$$

The latter expresses the dielectric constant in terms of tuning or detuning between two oscillations. When $\omega = 0$ (that is, there is no oscillating field), then the latter relation reduces to the static dielectric constant, $\varepsilon > 1$. When $\omega > \omega_0$, then it reduces to a dielectric constant, $\varepsilon < 1$. However, according to theory, for a frequency region $\omega > \omega_0$ and under the condition $\omega^2 < \omega_0^2 + 4\pi Ne^2/m$, the second term is such that $\varepsilon(\omega)$ becomes negative. If such conditions are achievable, that is $\omega > \omega_0$ and $\omega_0^2 < 4\pi Ne^2/3m$ or $\omega^2 < 16\pi Ne^2/3m$ (assuming $\omega = 2\omega_0$), then the *dielectric constant is negative,* the *refractive index is purely imaginary,* the *group velocity meaningless,* and a new chapter in photonics and optical communications opens.

2.11 MATERIAL DISPERSION

The refractive index of matter is related to the dielectric coefficient and to the characteristic resonance frequency of its dipoles. These dipoles interact with optical frequencies that are in the neighborhood of the resonant frequency. The closer they are, the stronger they interact and the more optical energy they absorb. Consequently, the refractive index becomes a function of the optical frequency ω, $n(\omega)$. However, the propagation constant is also related to the refractive index and, as a result, different optical frequencies propagate at different velocities. This velocity variability causes dispersion, termed *material dispersion,* and over the length of fiber it is measured in ps/nm. Dispersion is more easily explained if one considers an optical signal that consists of a narrow band of frequencies. Then, dispersion represents a time delay difference per nanometer of frequency, spectrally distributed in an optical signal. That is, a 100 nsecond delay over a 10-frequency spectrum is different than a 20 nsec delay over the same spectrum.

Silica, a key ingredient of optical fiber and cable, has a refractive index that varies with optical frequency. Thus, dispersion plays a significant role in fiber-optic communications.

The polarization of an electromagnetic wave, P, induced in the electric dipoles of a medium by an electric field, E, is proportional to susceptibility, χ:

$$P = \varepsilon_0[\chi^1 \cdot E + \chi^2 \cdot E \cdot E + \chi^3 \cdot E \cdot E \cdot E + \ldots]$$

where ε_0 is the permitivity of free space.

For an isotropic medium, the first-order term expresses the linear behavior of matter. The second-order term is orthogonal and thus it vanishes. For anisotropic media, the second order still holds. Higher-order terms are negligible (in optical communications, this may not be true for long fiber lengths and for ultrahigh bit rates).

For an isotropic medium, the nonlinear terms vanish and the above series relation is simplified to $P = \varepsilon_0 \chi \cdot E$. However, nature is not as simple, and most materials are either anisotropic, or become anisotropic under certain conditions. In such cases, higher-order terms should also be considered. In particular, the third-order term becomes significant and it results in nonlinear effects that may affect and limit optical transmission.

The most influential nonlinear effects in optical transmission, particularly when many wavelengths at high optical power are transmitted over the same medium (e.g., DWDM), are *four-wave mixing* (FWM), *stimulated Raman scattering* (SRC), and *Brillouin scattering* (BS).

2.12 GLASS FIBER, AN OPTICAL TRANSMISSION MEDIUM

Electromagnetic waveguides have been the subject of study in conducting electromagnetic waves since the emergence of microwave transmission. These waveguides have a rectangular or circular cross section, are hollow, and are millimeters or centimeters wide and several meters long. A single fiber strand is an electromagnetic circular waveguide filled with a dielectric, micrometers wide and many kilometers long. Fiber consists of ultrapure silica glass (SiO_2, 70–95 wt%) doped with specific elements such as germanium, fluorine, phosphorus, and boron, known as *dopants*. Dopants are added to adjust the distribution of the *refractive index profile* and light-propagation characteristics. The purity of glass fiber is measured by its very low loss, about 0.35 dB/km at 1310 nm and 0.25 dB at 1550 nm.

The optical fiber consists of two concentric layers. The innermost layer is the silica *core* in which the photonic signal propagates. The core is surrounded by another layer of silica known as *cladding*. The cladding has a lower refractive index than the core. Some parameters that pertain to fibers are:

- Fiber type (multimode, single mode, dispersion compensating)
- Forward attenuation per kilometer
- Backward attenuation per kilometer (if asymmetric)
- Polarization mode dispersion
- Polarization-dependent loss (PDL)
- Birefringence
- Dispersion (chromatic and material)
- Zero-dispersion wavelength
- Dispersion flatness over spectrum range
- Cut-off wavelength

2.12.1 Fiber Modes

When monochromatic light travels in optical waveguides, transmission takes place through specific guided modes. These modes are determined from the *eigenvalues* of second-order differential equations and their boundary conditions, similar to the propagation of electromagnetic waves in cylindrical waveguides. The solution of these equations determines the modes of propagation in the waveguide, as well as the *cut-off frequency* beyond which the fiber does not support transmission.

The propagation characteristics of light in silica fiber depend on the chemical consistency (silica plus dopants) and the cross-sectional dimensions of core and cladding. Typically, core plus cladding have a diameter of about 125 μm, but the core itself comes in two sizes, depending on the application the fiber is intended for. Fiber with a core diameter of about 50 μm (or about 62.5 μm) supports many modes of propagation and thus is known as *multimode* fiber (Figure 2.9). Fiber with

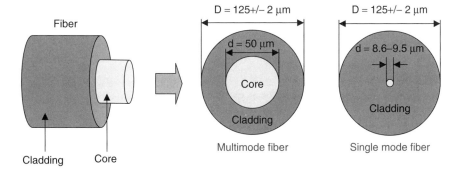

Figure 2.9. Geometric differences between multimode and single-mode fiber.

a core diameter between 8.6 and 9.5 μm supports one mode of propagation and is known as *single mode* (per ITU-T Recommendation G.652).

2.12.2 Multimode Fiber

Multimode and single-mode fibers have different refractive index profiles, different cross-sectional dimensions, and, therefore, different transmission characteristics. Consequently, different fiber types are used in different network applications, such as multimode in access and single mode in long haul.

Multimode fiber has various radial distributions of refractive index. Two well-known refractive indexes are the graded index and the step index. The salient characteristics of the multimode-fiber graded index (MMF GRIN) are:

- It minimizes delay spread, but delay is still significant in long lengths.
- A 1% index difference between core and cladding amounts to 1–5 ns/km delay spread (compare with the step index, which has a spread of about 50 ns/km).
- Easier to splice and easier to couple light into it.
- Bit rate is limited: up to 100 Mbit/s for lengths up to 40 km. Higher bit rates for shorter lengths.
- Fiber span without amplification is limited—up to 40 km at 100 Mbit/s (extended to Gbps for shorter distances for graded index).
- Dispersion effects for long lengths and high bit rates is a limiting factor.

The actual number of modes, M, that a fiber supports depends on fiber geometry (core diameter), the refractive index of core and cladding (or the numerical aperture), and the wavelength of the optical signal. These fiber parameters are combined in a normalized parameter known as the M number:

$$M = \tfrac{1}{2}[(4\pi/\lambda)d\sqrt{n_{\text{clad}}^2 - n_{\text{core}}^2}]^2$$

where λ is the wavelength and d is the core diameter.

The square root in the latter is known as the *numerical aperture* (NA) of a step index fiber:

$$NA = \sqrt{n_{\text{clad}}^2 - n_{\text{core}}^2}$$

2.12.3 Single-Mode Characteristics

Similarly, the salient characteristics of standard single mode fiber (SSMF) are:

- It (almost) eliminates delay spread.
- More difficult to splice (than multimode) due to critical core alignment requirements.

- More difficult to couple all photonic energy from a source into it.
- Difficult to study propagation with ray theory; it requires Maxwell's equations.
- Suitable for transmitting modulated signals at 40 Gbit/s (or higher) and up to 200 km without amplification.
- Long lengths and bit rates greater than 10 Gbit/s raise a number of issues due to residual nonlinearity/birefringence of the fiber.
- Fiber temperature for long lengths and bit rates greater than 10 Gbit/s becomes significant.

2.12.4 Effective Area

Consider the cross section of a fiber with a step index; that is, the refractive index is constant along the radius of the cross section of the core. Thus, when light propagates along the fiber core, photonic energy is distributed uniformly across the core, and for all practical cases, when the area of the core is needed to calculate power density or some figure of merit, the area is calculated as $A = \pi r^2$. However, in most fibers, the distribution of the refractive index is not constant. Although the physical dimensions of the fiber core remain the same (9 μm with a physical area $A = 63.5$ μm^2), the area experienced by the photonic energy (at a specific wavelength) is slightly different. This area is termed the "effective area" (A_{eff}) and this is used instead.

The effective area depends on the physical dimension of the core, on the distribution of the refractive index along its radius, and on the wavelength. Therefore, there is no straightforward formula for all fibers and cases and, as a consequence, an estimation of A_{eff} is provided by fiber manufacturers. Typically, the effective area may vary from 40 to 80 μm^2; in specialty fibers, A_{eff} may higher than that (Table 2.3). A ratio that is also of interest in many calculations involves A_{eff} and the

Table 2.3. Parameters, including the effective area for certain fibers

Fiber	D (ps/nm-Km)	S (ps/nm^2-Km)	A_{eff} (μm^2)	Loss (dB/Km)
TWRS	4.4	0.045	55	0.21
TWRS+	4.4	0.030	55	0.21
LEAF	4.0	0.09	72	0.21
SSMF	17	0.056	80	0.21
TeraLight	8	0.058	65	0.21
DCF	−100	−0.33	20	0.5

D = chromatic dispersion coefficient
S = dispersion slobe coefficient
A_{eff} = fiber core effective area

nonlinear refractive index of the fiber n_2 is n_2/A_{eff}; this ratio is termed the non-linear coefficient (ITU-T G.655). An implication that is of interest in this book is that as the effective area of the fiber increases the probability of having bits in error reduces. Thus, in general, it is preferable to use fibers with larger A_{eff}.

2.12.5 Cut-off Wavelength

In optical fibers with complex refractive index profiles, the theoretical calculation of numerical aperture and cut-off wavelength is more complex and it is determined with approximate models. Per ITU-T G.650, the cut-off wavelength is defined as that wavelength that experiences the waveguide a "fundamental mode power decrease to less than 0.1 dB". ITU-T G.650 also elaborates that in the case when many modes are equally excited, "the second-order (LP11) mode undergoes 19.3 dB more attenuation than the fundamental (LP01)" mode.

In general, if the wavelength in free space is λ and the wavelength in the guide is λ_g, then the cut-off wavelength λ_c is

$$(1/\lambda_c^2) = (1/\lambda^2) - (1/\lambda_g^2)$$

In terms of the M number, the condition for single-mode transmission is approximated to

$$M \leqq M_{cutoff} = \sim 2.4$$

Similarly, the wavelength below which a single-mode fiber allows multiple modes is

$$\lambda \geqq \lambda_{cutoff} = \sim 2.6r(NA)$$

where r is the core radius and NA is the numerical aperture of the fiber.

2.12.6 Fiber Attenuation and Power Loss

Fiber consists of optically transparent matter that, like any other matter, absorbs and scatters part of the optical power of light due to impurities in it, imperfections, and refractive index fluctuations. The total optical power lost in fiber is termed *attenuation*. However, power attenuation is also a function of the propagating frequency.

There are also additional mechanisms that contribute to optical power loss. The sum of all losses is clearly measured by subtracting the power out from the power of the fiber. This is termed *fiber loss* (many times used interchangeably with power attenuation). Fiber loss, for a given optical power $P(0)$ launched into the fiber, affects the total power arrived at the receiver, P_r. Based on this, fiber loss limits the fiber span, L_{max}, without amplification, and/or determines the required amplification gain.

2.12 GLASS FIBER, AN OPTICAL TRANSMISSION MEDIUM

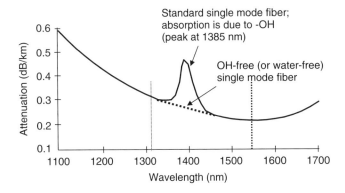

Figure 2.10. Single-mode fiber attenuation as a function of wavelength.

For a fiber with optical power attenuation constant $\alpha(\lambda)$, the optical power attenuation at a length L is expressed as

$$P(L) = P(0) \times 10^{-\alpha(\lambda)L/10}$$

In this relationship, if we replace $P(L)$ with the minimum acceptable power at the receiver, P_r, then the (ideal) maximum fiber length is

$$L_{max} = [10/\alpha(\lambda)]\log_{10}[P(0)/P_r]$$

In subsequent sections, we will see that there are additional limiting factors (e.g. dispersion and bit rate) that further limit the (ideal) maximum fiber length.

In general, the optical power attenuation constant, $\alpha(\lambda)$, is nonlinear and depends on the wavelength:

$$\alpha(\lambda) = C_1/\lambda^4 + C_2 + A(\lambda)$$

where C_1 is a constant (due to Rayleigh scattering), C_2 is a constant due to fiber imperfections, and $A(\lambda)$ is a function that describes fiber-impurity absorption as a function of wavelength.

The optical power attenuation constant of a fiber (measured in dB/km) is typically plotted as a function of the wavelength (Figure 2.10).

Conventional single-mode fibers have two low attenuation ranges—one at about 1.3 μm and another at about 1.55 μm. Between these two and at about 1.4 μm there is a high-attenuation range (1350–1450 nm with peak at 1385 nm) due to the OH radical. ITU-T G.652 recommends losses less than 0.5 dB/km in the region 1310 nm, and less than 0.4 dB/km in the region 1500 nm.*

---

*Recently, low-water-peak fiber has been produced that allows the full spectrum from under 1300 to over 1600 nm to be utilized.

2.12.7 Limits of Optical Power in Fiber

The maximum acceptable optical power density is the amount of optical power that a fiber can support without being damaged. Power density is the ratio of laser beam power over the cross-sectional area of the laser beam. Because the cross-sectional areas are very small, low powers may have large densities. For example, a 10 mW power beam with a 6 μm diameter produces a power density of

$$P/A = 10 \times 10^{-3}/\pi(3 \times 10^{-4})^2 \text{ W/cm}^2$$
$$= 10 \times 10^{-3}/28.26 \times 10^{-8} = 35{,}000 \text{ W/cm}^2$$

In communications, the optical signal is modulated in the form of pulses. A pulse is typically characterized by duration (in psec or nsec) and energy per pulse (in millijoules). Thus, a rectangular pulse of 10 pJ for 1 nsec has an equivalent power of

$$P = 10 \times 10^{-12}/10^{-9} \text{ J/s} = 10 \text{ mW}$$

As a point of reference, 1 W = 1 J/s, 1 eV = 1.6×10^{-19} J, and the energy of one photon at 194 THz (in the C-band) is

$$E = h\nu = 6.6 \times 10^{-34} \times 194 \times 10^{12} \text{ J} \times \text{sec/sec} = 1280.5 \times 10^{-22} \text{ J} = {\sim}1.3 \times 10^{-7} \text{ pJ}$$

The shape of pulse and its duration are closely linked with modulation, bit rate and filtering (Figure 2.11). Having calculated the power in a pulse, or the instantaneous power, the remaining task is to calculate the power in a signal, or average power. To explain this, consider the simple case of on–off keying (OOK) NRZ modulation. Ideally, "one" symbols have energy, whereas "zero" symbols do not. However, in a very long string of bits, the number of "ones" equal the number of "zeros" as the probability of having a "one" or a "zero" is 0.5. Thus, the overall power of a very large number of symbols is the average of the power in both symbols, $P/2$. Thus, similar reasoning can be extended to OOK RZ pulses, or pulses with a Gaussian distribution, and so on.

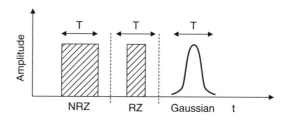

Figure 2.11. Three pulse shapes that illustrate the difference in area under the pulse. T denotes the period of a bit.

A pulse is characterized by duration (in psec or nsec) and energy per pulse in millijoules. In such cases, to calculate the power density of the pulse, the energy over time must first be converted to power and then to power density. Thus, a pulse of 10 mJ for 1 nsec is equivalent to a power of

$$P = 10 \times 10^{-3}/10^{-9} \text{ J/s} = 10 \text{ MW}$$

The maximum acceptable optical power density is the amount of optical power that a particular fiber type can support without being damaged. Power density is the ratio of laser-beam power over the cross-sectional area of the laser beam. Because the cross-sectional area is very small, even a low optical power may have a large density. For example, power density of a 10 mW beam focused on the disk of 10 mm diameter is

$$P/A = 10 \times 10^{-3}/\pi(5 \times 10^{-4})^2 \text{ W/cm}^2 = \sim 13 \text{ kW/cm}^2$$

Now, this is a continuous laser beam impinging on the disc that actually may be the surface of a fiber core. When the beam is modulated, a distinction must be made between the continuous case and the modulated beam that represents a dynamic power density. However, in either case, extremely low power levels yield extremely high power density and, depending on application, this is what makes it beneficial or destructive.

2.12.8 Fiber Birefringence

The assumption in ideal single-mode fibers is that there is no birefringence. Realistically, however, fiber has a small residual birefringence, which in some applications may be considerable, particularly as the span and/or the bit rate increases. The *degree of fiber birefringence* is defined by

$$B = |n_x - n_y|$$

where n_x and n_y are indices for the polarized fiber modes in the axes x and y (z is the axis of propagation).

Fiber birefringence causes power to be exchanged between the two polarization states in a continuous manner, changing polarization from linear, to elliptical, to circular, and, finally, back to linear. The length of birefringent fiber over which a complete revolution of polarization takes place is defined as *beat length,* which is

$$L_B = \lambda/B$$

For $\lambda = 1550$ nm and $B \sim 10^{-7}$, L_B is ~15 meters.

In coherent communications where a specific polarization at the receiver is expected, or when polarization-sensitive components are used, birefringence may be of concern. To minimize the effects of fiber birefringence, *polarization-preserving fibers* (PPF) may be used, which exhibit a very strong degree of birefringence ($B \sim 10^{-4}$). Thus, when the signal enters, the PPF the birefringence induced by the fiber is so strong than it "overshadows" other birefringence sources.

2.13 DISPERSION

Dispersion is the effect of widening the temporal and spectral content of a signal. In optical communications where pulses are subnanosecond, dispersion becomes very critical to the signal quality and the signal performance at the receiver. In digital transmission, a rule of thumb for acceptable dispersion (of any type) is

$$\Delta \tau < T/k$$

and the bit-rate information limit is expressed by

$$R_b < 1/(k\Delta \tau)$$

where $\Delta \tau = \tau_2 - \tau_1$, R_b is the information (bit) rate, T is the bit period, k is the dispersion factor (a transmission design parameter, typically selected to $k = 4$). If $k = 5$, then less dispersion is acceptable, and if $k = 3$, more is.

There are several mechanisms that cause signal dispersion. Dispersion is caused by signal imperfections and by the geometry and nonlinearity of the transmission medium. For example, the signal is not monochromatic but polychromatic, and the medium has geometrical and refractive-index variations. In addition, the latter exhibits nonlinear behavior and interacts with photons.

2.13.1 Modal Dispersion

An optical signal launched into fiber may be viewed as a bundle of nonparallel rays transmitted within a small cone. Because the rays are not parallel, they travel along different paths in the fiber in a zigzag manner, as electromagnetic waves do in waveguides; each zigzag path represents a different *mode* (Figure 2.12). For simpli-

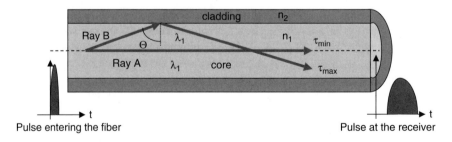

Figure 2.12. Modal dispersion causes a narrow pulse to widen.

fication of this description, we assume ideal monochromaticity. Thus, ray A travels in a straight path along the core of the fiber (one mode), whereas ray B travels at an angle, bouncing off the cladding (another mode). Thus, as rays at different modes travel different distances, they arrive at the receiver at different times. As a result, the initial narrow pulse spreads out. This is known as *modal dispersion* and it is more pronounced in multimode fibers.

The min–max travel time of two rays is expressed as

$$\tau_{min} = (Ln_1)/v$$

and

$$\tau_{max} = (Ln_1)/v \cos \Theta$$

for $\Theta = \Theta_{crit}$ (total reflection), and from Snell's law,

$$\cos \Theta_{crit} = n_1/n_2$$

The difference in travel time (assuming total reflection) is

$$\Delta \tau = \tau_{max} - \tau_{min} = [(Ln_1)/v]\Delta n/n_1$$

Hence, the maximum bit rate, R_b, is calculated from

$$R_b < 1/(4\Delta\tau) = (1/4)(v/Ln_1)(n_1/\Delta n)$$

Now, consider the case of two connected multimode fibers. Clearly, the connection point of the two fibers represents a perturbation or a discontinuity in the optical path. When light rays reach the end of the first fiber, they are launched into the second. However, since there are many modes, that is, different rays travel in different angles, each ray enters the second fiber at a different angle and thus is refracted differently, changing from one mode to another mode. This is known as *mode mixing* and it occurs in interconnected multimode fibers only.

Mode mixing affects the actual transmitted bandwidth (BW_{act}) over the multimode fiber length (L), defining an effective bandwidth. To distinguish the effect of mode mixing on bandwidth, an empirical *scaling factor*, γ, has been devised, defining the effective bandwidth as

$$BW_{eff} = BW_{act}/L^\gamma$$

where γ is between 0.7 and 1.0

The difference in travel time (assuming total reflection) is improved if a graded-index fiber is used, with a graded-index refraction profile $n(r)$. In this case, the fiber dispersion is improved if this condition holds:

$$R_b \leqq 2v_g/(n_g L \Delta n^2)$$

where n_g is the group refractive index, Δn is the maximum relative index between core and cladding, and v_g is the group velocity in the medium.

2.13.2 Chromatic Dispersion

The refractive index and the propagation constant of the fiber medium are functions of the wavelength. Thus, assuming no modal dispersion, the travel time of each wavelength over a unit length of fiber depends on these two parameters. Thus, as a pulse travels in the fiber, each wavelength travels at different speed and the pulse broadens. This is termed *chromatic dispersion* (Figure 2.13).

Therefore, chromatic dispersion is the rate of change of the group delay with wavelength. The general propagation constant β may be expanded in a Taylor series:

$$\beta(\omega) = \beta_0 + \beta_1(\omega - \omega_0) + 0.5\beta_2(\omega - \omega_0)^2 + \ldots$$

In the series, the term β_1 is the inverse of the group velocity, and β_2 is the group velocity dispersion, which causes pulse broadening.

In general, the photonic impulse response, $H(t)$, of a dispersive material can be expressed by

$$H(t) = \exp[(i\pi t^2)/(2\beta''z)]$$

where z is the optical path distance in the dispersive medium, and β'' is the second derivative of the propagation constant with respect to frequency ω (or the second-order dispersion coefficient).

Chromatic dispersion consists of two contributions, *material dispersion* and *wavelength dispersion,* and by some sources is also termed *waveguide dispersion.* Dispersion is measured in ps/nm-km (i.e., delay per wavelength variation and per fiber length).

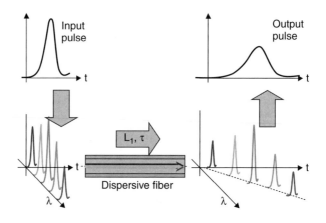

Figure 2.13. Fundamentals of chromatic dispersion.

Material dispersion is due to the dependence of the dielectric constant, ε, or the refractive index, n, on frequency, ω. Thus, the propagation characteristics of each wavelength in a fiber are different. Different wavelengths travel at different speeds in the fiber, which results in dispersion due to the material.

Material dispersion is most significant and is characterized by a parameter M, defined as the derivative of group index, N, with respect to wavelength, λ:

$$M(\lambda) = -(1/c)(dN/d\lambda) = (\lambda/c)(d^2n/d\lambda^2)$$

where n is the refractive index, λ is the wavelength, c is the speed of light in free space, and $N = n - \lambda(dn/d\lambda)$.

Consider a medium with a region in which the distribution of n changes varies nonlinearly; then the group and phase velocity change accordingly. Now, if a narrow light pulse that consists of a narrow range of wavelengths is launched in a medium, each individual wavelength arrives at the end of the fiber at different time. The result is a dispersed pulse due to *material dispersion.*

Waveguide dispersion or *wavelength dispersion* is the contribution due to nonlinear dependence of the propagation constant on frequency, ω. Wavelength dispersion is explained as follows.

Assume a narrow optical impulse that consists of a narrow spectral range. Consider two wavelengths λ_1 and λ_2 in the same impulse. We assume that both wavelengths travel (along the core of the fiber) in a straight path, but λ_1 travels faster than λ_2 ($\lambda_1 < \lambda_2$) due to the nonlinear dependence of the propagation constant on frequency ω (and on wavelength).

Wavelength dispersion has a different sign than material dispersion (Figure 2.14), although material dispersion is the major contributor. The counteracting action of wavelength dispersion to material dispersion slightly ameliorates the com-

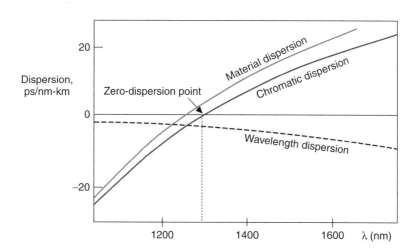

Figure 2.14. Two contributors to chromatic dispersion: material and wavelength.

bined effect. Negative dispersion implies that shorter wavelengths travel slower than longer wavelengths. As a consequence, fiber with negative dispersion can be used to compensate for positive dispersion (see Section 2.13.x).

The travel time t for a group velocity v_g over a length of fiber L is

$$\tau = L/v_g$$

or

$$\tau = L\beta' = L(\partial\beta/\partial\omega)$$

where β is the propagation constant, β' is the first derivative with respect to ω, and ∂ is the partial derivative.

The variation of τ with respect to ω, $(\partial\tau/\partial\omega)$, is

$$\partial\tau/\partial\omega = L\partial(1/v_g)/\partial\omega = L\partial^2(\beta)/\partial\omega^2 = L\beta''$$

where β'' is the second derivative with respect to ω.

For a signal with a spectral width $\Delta\omega$,

$$\Delta\tau = (\beta'')L\Delta\omega$$

That is, the pulse spread $\Delta\tau$ (chromatic dispersion) depends on the second derivative with respect to ω, β'', and it is proportional to the length of the fiber, L, and the spectral range $\Delta\omega$.

Based on this, a *group velocity dispersion* (GVD) coefficient, D, is defined as the variation of travel time due to the wavelength variation per unit length of fiber, L:

$$D = (1/L)(\partial\tau/\partial\lambda)$$

The coefficient D is also known as *chromatic dispersion coefficient* and it gives a measure of the *group delay rate change* with wavelength. Consequently, in communications, the chromatic dispersion of a fiber can be found by measuring the time delay of wavelength components that constitute an optical pulse.

It follows that

$$D = (1/L)(\partial\tau/\partial\omega)(\partial\omega/\partial\lambda)$$

However,

$$\partial\tau/\partial\omega = L\beta''$$

and

$$\partial\omega/\partial\lambda = -2\pi v/\lambda^2$$

and thus

$$D = -(2\pi v/\lambda^2)\beta''$$

and

$$\Delta\tau = DL\{[-1/(2\pi v/\lambda^2)]\Delta\omega\}$$

Finally, the pulse spread, or chromatic dispersion, is expressed by (∂ has been replaced by Δ):

$$\Delta\tau = |D|L\Delta\lambda$$

where $\Delta\lambda$ is the optical spectral width of the signal (in nm units).

Clearly, the pulse spread due to dispersion imposes a limitation on the maximum bit rate, known as the dispersion penalty, which is measured in dB. The dispersion penalty is related to the maximum allowable delay (a fraction ε of the bit period T) before severe signal degradation and unacceptable bit error rate (BER) occur. A 1 dB dispersion penalty corresponds to a bit-period fraction of approximately 0.5. At ultrahigh bit rates, 0.5 T may be a few picoseconds.

2.13.4 Dispersion-Shifted Fiber

The material dispersion parameter is given by:

$$M(\lambda) = -(dN/d\lambda)/c = (\lambda/c)(d^2n/d\lambda^2)$$

Because of the nonlinear dependency of the refractive index, at some wavelength the derivative $d^2[n(\omega)]/d\lambda^2$ becomes zero. The value λ_0 for which M becomes zero is known as *zero-dispersion wavelength*. Thus, M is positive for $\lambda > \lambda_0$ and negative for $\lambda < \lambda_0$. In certain cases, the wavelength dispersion and the chromatic dispersion within a spectral (wavelength) range may be of opposite signs and, thus, minimize the net effect. The wavelength range over which this may take place depends on the dispersion slope, given by

$$dD(\lambda)/d\lambda = \sim(\lambda/c)(d^3n/d\lambda^3)$$

In general, the operating point on the dispersion curve is not at zero wavelength but a small dispersion is desirable to minimize nonlinear interactions that result in pulse-shape distortion and, thus, noise generation. For example, when the laser source is expected to increase in frequency, such as due to temperature rise (a phenomenon known as *positive chirping*), then the operating point is set below the zero wavelength; that is, a small negative dispersion is desirable *as it starts at the outset with pulse compression.*

The amount of desirable dispersion is a function of data rate and modulation method (return-to-zero versus nonreturn-to-zero). As data rate increases, the pulse width narrows and pulse distortions are more pronounced. Similarly, a decreased optical power density (power per cm^2 of fiber core cross section) reduces the nonlinear effects, and, thus, an increased effective area of the fiber core is also desirable. A conventional single-mode fiber with a core diameter of about 8.3 μm and an index of refraction variation of about 0.37% has zero dispersion at about 1.3 μm. Below this, wavelength dispersion is negative and above it it is positive.

Dispersion-shifted fiber (DSF) is fiber with the zero-dispersion point shifted to 1550 nm (1.55 μm), that is, where the minimum absorption for silica fiber.

2.13.5 Dispersion Slope and Dispersion Compensation

Chromatic dispersion is compensated for using a *dispersion-compensating fiber* (DCF). A DCF is a fiber that causes dispersion opposite to conventional single-mode fibers. For instance, if SMF has positive dispersion over a specific range of wavelengths, DCF has negative dispersion over the same range.

Typically, the core of DCF is heavily doped with germanium and the cladding with fluoride, and with a proper index profile, a negative dispersion is achieved. In addition, the DCFs core has an effective aperture three to four times greater and a refractive index about five times that of standard SMF, but the DCF attenuation coefficient is twice that of standard SMF. Thus, if standard SMF and DCF fiber are alternated over the complete length, the dispersion induced by the SMF is undone by the DCF (Figure 2.15).

However, SMF and DCF are not spectrally balanced and if there is more than one optical channel in a fiber, as in DWDM, all optical channels are not compensated to the same degree, which results in a residual dispersion. The variation of residual dispersion is known as *dispersion slope mismatch* (Figure 2.16).

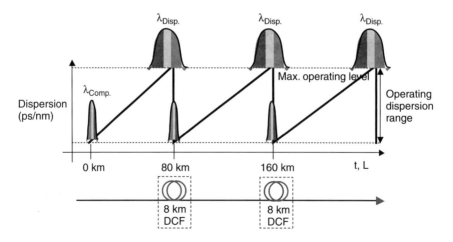

Figure 2.15. Dispersion compensation over a long span.

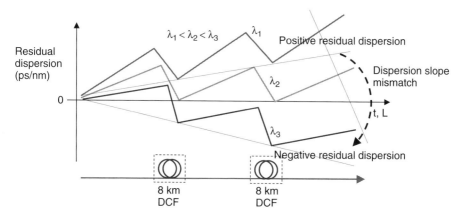

Figure 2.16. DCF is not spectrally balanced, resulting in residual dispersion.

2.13.6 Polarization-Mode Dispersion

All fibers have some residual birefringence and a core that is not perfectly circular over their entire length. Fiber birefringence and core noncircularity cause an optical (monochromatic) signal to be separated into two orthogonally polarized signals, or principal states of polarization (PSP), each traveling at a different speed and phase. The same thing happens to each pulse of a modulated optical signal; the pulse is separated into two pulses, each traveling at a different speed. Thus, when the two signals recombine, because of the variation in time of arrival, a pulse spreading occurs. This phenomenon is particularly noticeable in single-mode fiber transmission at ultrahigh bit rates (above 2.5 Gbps) and is known as *polarization-mode dispersion* (PMD).

By definition (see ITU-T G.650), PMD is measured by the average differential group delay time (DGD) over wavelength between two orthogonally polarized modes (measured in ps). PMD is maximized if both PSPs are equally and maximally excited, resulting in maximum DGD and pulse spreading. Conversely, PMD vanishes or is greatly minimized if only one of the two states is excited. Similarly, the *polarization mode dispersion coefficient* is defined as the PMD divided by the square root of fiber length (measured in ps/$\sqrt{\text{km}}$). Optical fibers have a polarization-mode dispersion coefficient of less than 0.5 ps/$\sqrt{\text{km}}$ (see ITU-T G.650, G.652, G.653, and G.655).

As a consequence of the DGD definition, if DGD is measured by the average difference of time arrival, $\Delta \tau$, between the two orthogonally polarized modes, and the two polarized modes are related to the birefringence of the fiber, Δn_g, then DGD is measured by

$$\Delta \tau = (\Delta n_g L)/c$$

where c is the speed of light, L is the length of the fiber, and n_g is the refractive index variation corresponding to the group velocity of the orthogonal polarization states.

2.13.7 Polarization-Mode Dispersion Compensation

Fibers are specified by an average *differential group delay* (DGD) in ps or a mean DGD coefficient in ps/$\sqrt{\text{km}}$. For a low-PMD fiber, the mean DGD coefficient is less than 0.1 ps/$\sqrt{\text{km}}$, and for a high-PMD fiber it is ~2 ps/$\sqrt{\text{km}}$. Thus, for a fiber length of 625 km the average DGD for low-PMD fiber is about 2.5 ps, whereas for high-PMD fiber it is about 50 ps ($2\sqrt{625} = 50$). Now, in terms of decibels, the receiver sensitivity penalty for an average DGD of 50 ps is less than 0.1 dB at 2.5 ps, and greater than 4 dB at 10 ps. Extrapolating to 40 ps, there is a severe receiver penalty. As consequence, PMD becomes a main contributor to dispersion at ultra-high bit rates, necessitating PMD compensation. However, compared with chromatic dispersion, PMD is much more difficult to compensate for.

2.13.8 Polarization-Mode Coupling

PMD has been phenomenologically explained by extensive vectorial analysis that considers the propagation of principal states of polarization. In a graphical representation of polarization states, known as the Poincaré sphere, PSPs are located at diametrically opposite points on the sphere's surface; points on the equator of the sphere represent linear polarization, polar points represent circular polarization, and points on the hemispheres represent elliptical polarization states (the northern hemisphere represents right-handed states and the southern left-handed states). As the noncircularity of a fiber core causes PSPs, each polarization state at different speed and phase, an issue emerges at the connecting point of two fibers where in most practical cases, core noncircularity is not matched. Hence, slow and fast states from one segment are coupled into another core with different orientations of polarization. Thus, at the interface, as the polarization states from one fiber are coupled into the next, input PSPs are transformed into different output PSPs. This is known as *polarization mode coupling* (PMC). As a consequence, over a fiber span with several connections, PMC becomes a random event that affects DGD and PSPs randomly to further complicate PMD compensation.

2.13.9 Pulse Timing and Dispersion

As optical pulses propagate in the fiber, the pulse characteristics degrade due to attenuation, noise, jitter, and nonlinear effects from photon–matter interactions. As a result, the timing pulse characteristics, rise time and fall time, degrade. If we consider that the total timing degradation of pulses ($\Delta \tau_{TOT}$) is the contribution of the transmitter ($\Delta \tau_{TR}$), fiber ($\Delta \tau_{FIB}$) and receiver ($\Delta \tau_{REC}$), then the following relationship holds:

$$\Delta \tau_{TOT}^2 = \Delta \tau_{TR}^2 + \Delta \tau_{FIB}^2 + \Delta \tau_{REC}^2$$

The contribution $\Delta \tau_{FIB}$ consists of all dispersion contributing mechanisms modal ($\Delta \tau_{MOD}$), chromatic ($\Delta \tau_{GVD}$), and polarization ($\Delta \tau_{POL}$), and the rise/fall time for the transmitter and receiver are obtained from device specification. That is,

$$\Delta\tau_{FIB}^2 = \Delta\tau_{MOD}^2 + \Delta\tau_{GVD}^2 + \Delta\tau_{POL}^2$$

In single-mode fiber, the modal dispersion may be considered negligible. In addition, for bit rates up 10 Gbit/s and moderate fiber lengths (<100 km), polarization dispersion may be considered negligible and, therefore, $\Delta\tau_{FIB}^2 = \Delta\tau_{GVD}^2$ and $\Delta\tau_{TOT}^2 = \Delta\tau_{TR}^2 + \Delta\tau_{GVD}^2 + \Delta\tau_{REC}^2$. Substituting for chromatic dispersion $\Delta\tau_{GVD}^2 = |D|L\Delta\lambda$, one obtains a relationship for the fiber length that meets the pulse timing requirements:

$$L = \frac{\sqrt{\Delta\tau_{TOT}^2 - \Delta\tau_{TR}^2 + \Delta\tau_{REC}^2}}{|D|\Delta\lambda}$$

Furthermore, if we consider the practical rule that the total rise and fall time degradation of the pulse at the receiver threshold level should not exceed half the pulse period 0.5 T, $\Delta\tau_{TOT} \leq 0.5\ T$ (for NRZ modulation), then the above relationship is further simplified as:

$$L \leq \frac{\sqrt{0.25\ T^2 - \Delta\tau_{TR}^2 + \Delta\tau_{REC}^2}}{|D|\Delta\lambda}$$

Clearly, if the limiting factors that we considered as negligible are not so, then the length of fiber to meet timing requirement decreases. Moreover, the fiber length is a function of wavelength since all limiting factors in the aforementioned relationship depend on wavelength. However, wavelength variability has been compensated for by the practical rule $\Delta\tau_{TOT} \leq 0.5\ T$.

2.14 FIBER POLARIZATION-DEPENDENT LOSS

Fiber, like most optically transparent materials, has a small degree of polarization sensitivity. That is, optical power passes through it without reduction in certain selective polarized sates but with reduced power in others. This is known as fiber *polarization-dependent loss* (PDL) and it represents the peak-to-peak optical power variation measured in decibels (dB). Linear polarizers have strong PDL (e.g., >30 dB) but very little SMF (<0.02 dB). Strong PDL adds to IL and cross talk.

PDL becomes critical when the bit rate is greater than 10 Gbit/s and the line width very narrow—at or less than 0.05 nm at full width half-maximum (FWHM). Very narrow linewidths become highly polarized. An issue with polarization dependence is that a source may change its polarization state randomly and thus its PDL, complicating power-budget calculations and degrading the quality of signal and system performance.

2.15 SELF-PHASE MODULATION

The dynamic characteristics of a propagating optical pulse in fiber, due to the Kerr effect of the medium, result in modulation of its own phase. This nonlinear phenomenon, known as *self-phase modulation,* causes spectral broadening. If the nonlinear refractive index of the fiber is n, the temporal variation of the electric field of the photonic signal causes the (nonlinear) Kerr effect. This deals with the variation of n in the time domain (dn/dt). Thus, as a photonic pulse propagates, a temporal variation of phase, Φ, takes place. According to electromagnetic wave theory, in a nonlinear medium, the phase derivative is proportional to frequency, $d\Phi/dt = 2\pi f$, which in this case is manifested by the generation of intense sidetones (undertones and overtones) that affect the spectral distribution of the signal and cause spectral broadening (Figure 2.17). Moreover, if the wavelength of the pulse is below the zero-dispersion point (known as a *normal dispersion regime*), then spectral broadening causes temporal broadening of the pulse as it propagates. If the wavelength is above the zero-dispersion wavelength of the fiber (the *anomalous dispersion regime*), then chromatic dispersion compensates for self-phase modulation and reduces the temporal broadening.

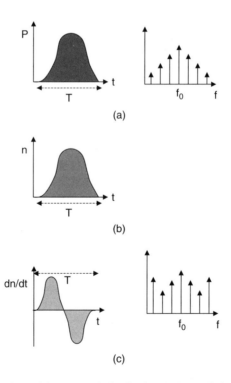

Figure 2.17. As a pulse (with a spectral distribution) (A) travels in a nonlinear medium, it affects its refractive index (B), the slope of which (C), according to the Kerr effect, causes spectral broadening and shape distortion.

2.16 SELF-MODULATION OR MODULATION INSTABILITY

When a single pulse of almost monochromatic light has a wavelength above the zero-dispersion wavelength of the fiber (known as the anomalous dispersion regime), another phenomenon occurs that degrades the width of the pulse. That is, two side lobes are symmetrically generated at either side of the pulse (Figure 2.18), thus adding to the noise content of the signal. This is known as *self-modulation* or *modulation instability*.

Modulation instability depends on material dispersion, the optical traveled path (or fiber length), and the optical channel power. Modulation instability is considered a special four-wave mixing case that affects the signal-to-noise ratio.

Modulation instability is reduced by operating at low energy levels and/or at wavelengths below the zero-dispersion wavelength.

2.17 EFFECT OF PULSE BROADENING ON BIT ERROR RATE

Pulse broadening is the combined effect of many degradation mechanisms, as already examined, that degrade the quality of the signal, generate intersymbol interference (ISI), and increase *bit error rate* (BER). Dispersion is one of them. As pulses broaden, they spread in adjacent bit periods (Figure 2.19). If in a bit period a logic "0" occurs, power spreading may raise the level such that the receiver may "see" it as a logic "1."

ITU has defined a number of fiber types suitable for different applications. For example, fiber with a high dispersion point (G.652), fiber with a large A_{eff} (G.655), and fiber with improved dispersion and dispersion slope (G.653).

2.18 FOUR-WAVE MIXING

Consider three optical frequencies, f_1, f_2, and f_3 closely spaced (in frequency). Then, from the interaction of the three, a fourth frequency is generated, f_{fwm}, such that f_{fwm}

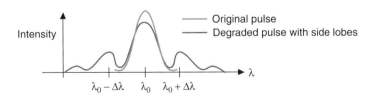

Figure 2.18. When an optical channel operates above the zero-dispersion point, two side lobes are symmetrically generated, a phenomenon known as self-modulation or modulation instability.

74 OPTICAL PROPAGATION

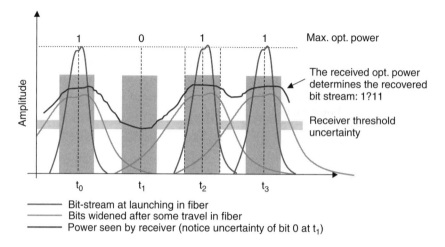

Figure 2.19. Effect of excessive pulse widening on ISI and BER.

$= f_1 + f_2 - f_2$. This is known as *four-wave mixing* (FWM), or *four-photon mixing* (Figure 2.20). FWM is a DWDM phenomenon.

The efficiency of FWM depends on:

- Channel spacing
- Power intensity of the contributing frequencies
- Phase matching of the contributing channels
- Chromatic dispersion of the fiber; because FWM performs best at the zero-dispersion wavelength, the operating point is preferably slightly above the zero-dispersion wavelength.
- Refractive index
- Fiber length
- Higher-order polarization properties of the material (nonlinear Kerr coefficient)

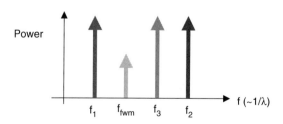

Figure 2.20. Three adjacent optical frequencies, f_1, f_2, and f_3, interact to produce a fourth frequency, f_{fwm}, where $f_{fwm} = f_1 + f_2 - f_3$, known as four-wave mixing.

FWM may also occur with two signals at different wavelengths, if their intensity and wavelengths are in a specific relationship. In such a case, the fiber refractive index is modulated at the beat frequency of the two wavelengths. In this case, the phase modulation creates two sidebands (at frequencies given by their difference) but at a lower intensity.

FWM affects the signal-to-noise ratio and thus the bit error rate. The effects of four-wave mixing on optical transmission and in single-mode fiber are manifested as:

- Cross talk, due to superposition of uncorrelated data from contributing channels; this is manifested as degradation of OSNR and OBER.
- Signal-power depletion as a result of signal-power sharing among the contributing channels to the FWM generated channel.
- Optical signal-to-noise degradation, due to superposition of noise and random data from the contributing frequencies.

As the signal input power of $f_1, f_2,$ and f_3 increases, or as the channel spacing decreases (due to denser channels in the spectral band), the FWM term increases. ITU-T (G.663) recommends that the critical optical power for FWM be greater than 10 mW, although this depends on channel spacing. At 200 GHz channel spacing, it has been experimentally verified that the FWM effect is drastically decreased as compared with 100 GHz or less spacing.

2.19 THE DECIBEL UNIT

Optical power amplitude, amplifier gain, power attenuation, optical loss, and power may be provided in units of decibels (dB). The decibel unit for power is defined as ten times the base-10 logarithm of power (in watts):

$$\text{power (dB)} = 10 \log_{10} (P \text{ in watts})$$

Implicitly, decibel conversions are computed with respect to a reference power unit in watts. In optical communications, the transmitted signal is on the order of milliwatts and, therefore, the dB units would always be large and negative numbers; for example, $10 \log (1 \text{ mW}) = 10 \log 10^{-3} = -30$ dBm. Therefore, the decibel unit has been modified to reflect small powers and a more practical unit for milliwatt power levels was developed—the dBm (pronounced dee-bee-em), which is defined as

$$\text{power (dBm)} = 10 \log_{10} (P \text{ in milliwatts})$$

When we deal with voltages, we need to modify the definition of dB. For example, from the relationship $\text{dB} = 10 \log [P/P_{\text{ref}}]$ and replacing P with V^2/R we have

$$10 \log [(V^2/R)/(V_{\text{ref}}^2/R)] = 10 \log (V/V_{\text{ref}})^2 = 20 \log (V/V_{\text{ref}})$$

Thus, dealing with voltages, the decibel level is calculated by multiplying the logarithm by 20 (and not by 10).

Because dB and dBm units are based on the properties of (base-10) logarithms, these properties are reviewed in the mathematical table that appears on page xxi.

When adding/subtracting dBs or dBms, certain cautions should be taken regarding the mix and match of units. Decibel units are additive if their argument is multiplicative. That is, if dB/dBm units are not handled correctly, one may make incorrect calculations.

Example 1
10 dBm = 10 log[10 mW], but 10 dBm + 10 dBm is *not* 20 dBm! If it were, then the result of 10 log(10 mW × 10 mW) would produce 100×10^{-6} W^2 = 10^{-4} W^2 and then one would calculate it as $10 \log 10^{-4}$ = –40 dB! In which case *the units are wrong* (W^2)

Example 2
3 mW yields 5 dBm and 8 mW yields 9 dBm, thus 3 mW + 8mW = 11 mW, which yields 10.5 dBm. Thus, if dBms could be added, then 5 dBm + 9 dBm would yield 14 dBm, which is wrong!

Example 3
x dBm + y dB = $(x+y)$dBm. Then, 5 dBm + 9 dB = 14 dBm, or 10 log 3 + 10 log 8. That is, we are adding:

$$10 \log [(3 \times 10^{-3})/(10^{-3})] + 10 \log(8) = 10 \log(3 \times 8) = 10 \log(24) = 14 \text{ dBm}$$

As a consequence:

- dBms *cannot* be added.
- dBms can be subtracted; the result is in dBs (not in dBms) since a dBm difference implies a power ratio.
- dBs can be added.
- A dB can be added to a dBm; the result is in dBm.

Although the absolute power of an optical signal in communications is in milliwatts and its attenuation or loss is expressed in dB units (and not in dBm). This is because attenuation or loss is the ratio of power-out over power-in, and, thus, the ratio becomes dimensionless. Hence, the attenuation coefficient, $\alpha(\lambda)$, per fiber km is expressed in dB:

$$\alpha(\lambda) = 10 \log P_{out}/P_{in} \text{ (dB)}$$

where P_{out} and P_{in} are measured with the same units (watts or milliwatts). Thus, the total fiber attenuation is the product of $\alpha(\lambda)xL$, where L is the fiber span in km. In the case of attenuation and loss, P_{out} is smaller than P_{in}, and the result is negative dBs. A similar relationship holds for amplification or gain, $g(\lambda)$, but in this case the result is positive dBs:

$$g(\lambda) = 10 \log P_{out}/P_{in} \text{ (dB)}$$

As an example, a power ratio of 1,000 is 30 dB, of 10 is 10 dB, of ~3 is 5 dB, of 2 is ~3 dB (these refer to gain), and a power ratio of 0.1 is –10 dB (this refers to loss). Notice that a ratio of 1 is 0 dB, as there is neither loss nor gain.

Power loss besides decibels is also provided as a percentage, such as 60% means that 60 power units were lost of 100 transmitted, and, thus, the power units received are $100 - 60 = 40$. Thus, the power transmittance in percent is $(100 - 60)/100 = 0.4$ or 40%. The correspondence of dB to percent is easy to calculate. For example, 90% power loss corresponds to $10 \log\{(100 - 90)/100\} = -10$ dB, 50% corresponds to $10 \log 0.5 = -3$ dB, and 2% to $10 \log 0.98 = -0.01$ dB. Table 2.4 lists conversions from dB loss to % loss and %loss to dB loss.

Example 4
Convert –1 dBm optical power in milliwatts. From the definition of dB,

$$-1 \text{ dBm} = 10 \log(x) = 10 \log(0.794 \text{ mW})$$

The optical power is

$$P = 0.794 \text{ mW or } \sim 0.8 \text{ mW}$$

Example 5
Convert –1 dBm of optical power launched in a fiber core with diameter $D = 10$ μm to optical power density (W/cm²). The cross-sectional area of fiber is:

$$A = (\pi/4)D^2 = (3.14/4)[10 \times 10 - 4]^2 \text{cm}^2 = 0.785 \times 10^{-6} \text{ cm}^2$$

The optical power density is

$$P/A = 0.794 \times 10^{-3} \text{W}/[0.785 \times 10^{-6}] \text{cm}^2 = \sim 1.01 \times 10^{-3} \text{ (W/cm}^2\text{)}$$

Table 2.4. Conversion of dB to %

dB loss	% loss	% loss	dB loss
0	0	0	0
–0.1	–2.3	–0.5	–0.02
–0.5	–10.9	–1	–0.04
–1	–20.6	–5	–0.22
–2	–36.9	–10	–0.46
–5	–68.4	–40	–2.22
–10	–90.0	–90	–10.00
–20	–99.0	–99	–20.00

Example 6

Calculate the optical power loss of a laser beam of −1 dBm launched in fiber. For simplicity, assume a circular beam with a step-intensity distribution and $D_1 = 12$ μm diameter, and a fiber with step index, $D_2 = 10$ μm core diameter, and 5% surface reflectivity.

There are two loss contributions: (a) due to cross-sectional area mismatch between beam and fiber core, and (b) due to reflectivity.

(a) The cross section of the beam is

$$A = (\pi/4)D_1^2 = (\pi/4)[12 \times 10 - 4]^2 \text{cm}^2 = (\pi/4)[1.44 \times 10^{-6}]\text{ cm}^2$$

Similarly, the cross section of the fiber core is $A = (\pi/4)D_2^2 = (\pi/4)[10^{-6}]$ cm². Thus, the part of optical power impinging on the core is calculated from

$$[D_2/D_1] \times P = [10^{-6}/1.44 \times 10^{-6}] \times 0.8 \text{ mW} = 0.56 \text{ mW}$$

(b) From the optical power impinging on the fiber core, 95% is coupled in and 5% is lost due to reflectivity. Thus, the power launched in the fiber is

$$0.95 \times 0.56 \text{ mW} = 0.528 \text{ mW}$$

The latter is expressed in dBm as follows:

$$10 \log(0.528 \text{ mW}) = -2.77 \text{ dBm}$$

REFERENCES

1. S.V. Kartalopoulos, "*DWDM: Networks, Devices and Technology,*" Wiley/IEEE Press, 2003.
2. S.V. Kartalopoulos, "*Introduction to DWDM Technology: Data in a Rainbow,*" IEEE Press, 2000.
3. I.P. Kaminow and T.L. Koch (Eds.), *Optical Fiber Communications IIIA; Optical Fiber Communications IIIB,* Academic Press, 1997.
4. S.V. Kartalopoulos, "Elastic Bandwidth," *IEEE Circuits and Devices Magazine, 18,* 1, 8–13, January 2002.
5. S.V Kartalopoulos, "On the Performance of Multiwavelength Optical Paths in High Capacity DWDM Optical Networks," *SPIE Optical Engineering,* September 2004.
6. S.V. Kartalopoulos and C. Qiao, "A Glass with Light, Please," Inaugural Editorial in IEEE Optical Communications, *IEEE Communications Magazine, Optical Communications Supplement,* February 2003, S4.
7. S.V. Kartalopoulos and D. Raftopoulos, "Interferometry for the Evaluation of the Dynamic Optical and Mechanical Properties of Transparent Materials," *Engineering Fracture Mechanics, 19,* 6, 993–1003, 1984.
8. D. Raftopoulos and S.V. Kartalopoulos, "Method For Determining Properties of Optically Isotropic and Anisotropic Materials," U.S. patent no. 4,195,929, April 1980.

9. D. Raftopoulos and S.V. Kartalopoulos, "Method For Determining Stress-Optical Constants of Optically Isotropic and Anisotropic Materials," U.S. patent no. 4,119,380, October 1980.
10. M. Bouhiyate and S.V. Kartalopoulos, "Performance of Optimized Chromatic Dispersion Compensated 10 Gbps DWDM Links," in *Proceedings of NFOEC-2003 Conference,* Orlando, FL, September. 6–10, 2003, pp. 1162–1167.
11. G. Bellotti, A. Bertaina, and S. Bigo, "Dependence of Self-Phase Modulation Impairments on Residual Dispersion in 10-Gb/s-Based Terrestrial Transmissions Using Standard Fiber," *IEEE Photon Technology Letters, 11,* 7, 824–826, July 1999.
12. D.M Rothnie and J.E Midwinter, "Improving Standard Fiber Performance by Positioning the Dispersion Compensating Fiber," *Electronic Letters, 32,* 20, 1907–1908, 1996.
13. N. Takato et al, "128-Channel Polarization-Insensitive Frequency-Selection-Switch Using High-Silica Waveguides on Si," *IEEE Photon. Technology Letters, 2,* 6, 441–443, 1990.
14. ITU-T Recommendation G.650, "Definition and Test Methods for the Relevant Parameters of Single-Mode Fibres," 1996.
15. ITU-T Recommendation G.652, "Characteristics of a Single-Mode Optical Fiber Cable," April 1997.
16. ITU-T Recommendation G.653, "Characteristics of a Dispersion-Shifted Single-Mode Optical Fiber Cable," April 1997.
17. ITU-T Recommendation G.655, "Characteristics of a Non-Zero Dispersion-Shifted Single-Mode Optical Fiber Cable," October 1996.
18. ITU-T Recommendation G.661, "Definition and Test Methods for the Relevant Generic Parameters of Optical Fiber Amplifiers," November 1996.
19. ITU-T Recommendation G.662, "Generic Characteristics of Optical Fiber Amplifier Devices and Sub-systems," July 1995.
20. ITU-T Recommendation G.663, "Application Related Aspects of Optical Fiber Amplifier Devices and Sub-systems," October 1996.

CHAPTER 3

Optical Transmitters and Receivers

3.1 INTRODUCTION

In optical transmission systems, there are three key elements: the transmitter (laser and modulator), the photodetector, and the optical transmission medium (the fiber). Typically, the detector is characterized by a level of sensitivity to impinging optical power. That is, the optical signal must be greater than the sensitivity and within certain limits. A too strong optical power oversaturates the detector (as the sun blinds us when looking into it) and a too weak power will be detected with many errors. The optical fiber is the information conduit but it is lossy, so the propagating optical signal experiences power loss. Therefore, the transmitter must provide enough optical power to the signal that enters the fiber to overcome loss and arrive at the photodetector above its sensitivity threshold. Now, this discussion is very simplistic when applied to the DWDM case as transmission becomes a complex matter. Laser sources and modulators are not so ideally simple, the transmitted signal is not as perfect, the transmission medium is complex, and the receiver is more complex. In this chapter, we examine the transmitter and the receiver, their parameters and their characteristics.

3.2 THE TRANSMITTER

In DWDM systems, the main function of the optical transmitter is to provide a modulated optical signal that complies with a set of specifications (as specified by standards and application; see Table 3.1).

The actual parameter nominal value and min–max variability depend on the application the transmitter is designed for. For example, the bit rate may be as low as several Mbit/s or as high as 40 Gbit/s. The signal modulation may be nonreturn to zero (NRZ) or return to zero (RZ). The output optical power may be specified for short fiber lengths or for long fiber length. Center frequency may be based on a 80 λ grid or 40 λ grid, and for different fiber types (SSMF, DSF, and MMF).

For example, the output power depends on the type of laser source (DFB, Fabry–Perot, VCSEL, etc.), design specifics of the laser device, bias, temperature, and the controlling circuitry in the transmitter package. Thus, a DFB laser may vary from 1 mW to more than 10 mW. Similarly, the extinction ratio r is a measure of

Table 3.1. Parameters specific to sources

- Output power
- Center wavelength λ_0
- Center frequency drift limits
- Spectral line width
- Channel spacing
- Cutoff λ (tunable sources)
- Beam profile, distribution, and uniformity
- TE/TM mode of photonic beam
- Modulation depth (modulated sources)
- Extinction ratio (modulated sources)
- Side-mode suppression ratio (modulated sources)
- Transient parameters (rise/fall time) (modulated sources)
- Bit rate accuracy and max–min range (modulated sources)
- Noise and floor noise
- Signal-to-noise ratio
- Wander and jitter and allowed limits
- Chirp
- State of polarization
- Polarization stability
- Beam-intensity distribution
- Acquisition time of new center wavelength (tunable sources)
- Accuracy of new acquired center wavelength (tunable sources)
- Dynamic spectral range (tunable sources)
- Dependency on bias
- Dependency on temperature
- Life span during which parameters are within an acceptable range

the relative power when a logic "one" is transmitted with respect to a logic "zero" (or pulse/no-pulse of light). Ideally, there is no light emitted for the duration of a logic "zero"; this would result in an infinite ratio. Thus, if the powers for the symbols "one" and "zero" are P_1 and P_0, respectively, the average power over a period of observation is $P_{avg} = (P_1 + P_0)/2$ and $P_1/P_0 = r$, then

$$P_1 = 2r\, P_{avg}/(r + 1)$$

and

$$P_0 = 2P_{avg}/(r + 1)$$

Typically, we would like to set $P_{avg} = P_1/2$, in which case $P_1 = 2P_{avg}$ and $P_0 = 2P_{avg}/r$. However, as noise is added to the optical signal, this condition is disturbed and the extinction ratio changes. A smaller r implies a smaller difference between "zero" and "one" levels, a smaller signal-to-noise ratio, and a larger receiver penalty.

In communications, a transmitter is neither a simple nor a single component. It consists of several components (contained in the same package), one of which is the laser source. Other components are bias and temperature stabilizers, and power and wavelength monitors.

3.2.1 Lasers

An examination of the optical beam generated by a laser device reveals that the beam is not as perfect as one would expect. For example, the beam is not monochromatic but contains many frequencies, albeit within a narrow spectrum or line width (Figure 3.1). Clearly, this affects the dispersion characteristics of the optical signal in the fiber.

The cross-sectional intensity of the beam is not uniform or Gaussian but its distribution is distorted. Its high-intensity area is not distributed at the center of the beam's cross section (Figure 3.2). The beam is not narrow and parallel over a long length, but over a span it narrows to a "neck" and then widens.

Moreover, the polarization of the beam is not uniform in all directions. The generated light interacts with the field of nearby atoms within the laser's active region and the modulator. This interaction affects the strength of the electric and/or magnetic field of light differently so that the beam may have circular, elliptical, or even linear polarization (Figure 3.3).

Finally, all the above as well as the output power of the modulated laser beam may fluctuate over time, which further adds to the engineering challenges of the optical path. Such impairments clearly complicate the coupling onto the fiber, as well as the study of beam propagation in a dielectric waveguide. Therefore, the limits and nominal values of the laser source's parameters are important in communications and have been the study of numerous standards.

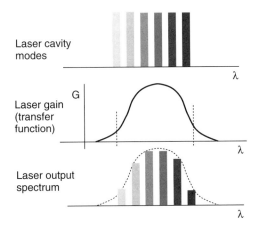

Figure 3.1. Multimode output of a laser cavity.

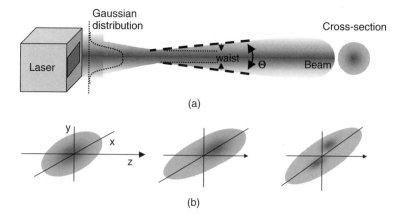

Figure 3.2. An ideal laser beam should be circular with a Gaussian intensity distribution (a). In reality, due to imperfections and parametric variations, the cross-sectional distribution of practical beams are elliptical and nonuniform (b).

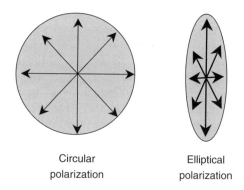

Figure 3.3. Two polarization modes, a circular and an elliptical. The direction of light is perpendicular to the page.

3.3 THE RECEIVER

The main function of the receiver is to detect the modulated photonic signal with an expected level of accuracy. The modulated photonic signal at the receiver has traveled along the fiber and through many components and it has been grossly attenuated in addition to being contaminated with noise and jitter (see Chapter 6). Thus, the incident light at the receiver is very weak and a logic "one" contains relatively few photons, which the photodetector must be capable of detecting reliably; the minimum number of photons required for a receiver to recognize a logic "one" is known as the *quantum limit,* which is rate dependent.

3.3 THE RECEIVER

A receiver consists of an optical preamplifier (optional); a polarization filter (optional); a power equalizer; a focusing lens (preferred); an efficient, fast-responding, and low-noise photodetector; an electronic low-pass filter; a transimpedance amplifier; a timing circuit; a threshold level; a digital pulse-shaping circuit; and a decoder (but not necessarily in this order).

In DWDM systems, after the optical preamplifier, polarizer, and equalizer, an optical wavelength demultiplexer separates the wavelengths, each of which is directed to its corresponding receiver. Thus, the key characteristics of a DWDM receiver are:

- Receiver sensitivity
- Optical preamplifier spectral response
- Optical preamplifier gain and gain flatness
- Optical preamplifier noise
- Demultiplexer polarization effects
- Power equalization spectral range and flatness
- Demultiplexer insertion loss
- Waveguide, connector, and splice insertion loss
- Waveguide polarization effects
- Detector technology (e.g., APD, PIN)
- Detector minimum threshold optical power, min–max threshold level (one–zero) at a given BER (e.g., less than 10^{-12})
- Detector quantum efficiency
- Detector dependency on polarization
- Detector temporal responsivity, max–min bit rate
- Detector spectral responsivity per wavelength, λ_0
- Detector wavelength discrimination
- Detector polarization dependency
- Detector shot noise
- Detector dependency on bias
- Detector dependency on jitter
- Detector dependency on temperature
- Low-pass filtering characteristics
- Receiver clock sensitivity to density
- Demodulation method
- Clock stability

The desired receiver sensitivity (R_s) is calculated from the power generated by the transmitter and coupled onto the fiber (T_x), minus the maximum allowable losses over the transmitter–receiver link ($Loss_{max}$), plus the optical gain G (if any) between transmitter and receiver:

$$R_s = T_x - \text{Loss}_{\max} + G \text{ (dB)}$$

Because in communications the ultimate goal is reliable delivery of data at the expected quality performance, what really matters is the ability of the receiver to detect and reconstruct the incoming signal to its initial form and within the performance parameters, which is expressed in number of errored bits received. Therefore, engineering a link must begin from the receiver and proceed toward the transmitter.

If the arriving signal at the receiver were ideal but attenuated, then the problem of receiving an almost error-free signal would be greatly simplified. However, this ideal case is far from real since the arriving signal may be heavily impaired by noise, frequency shift, polarization-state change, crosstalk, jitter and wander, dispersion, and reflections, in addition to attenuation (see also Chapter 5). In fact, these effects are specified and are included in the calculation of the optical-path penalty (OPP). Thus, the path penalty is the combined distortion of the signal waveform during its transmission over the fiber path that affects receiver sensitivity by increasing the bit error rate. Because in optical communications a performance objective is a 10^{-12} bit error rate, the path penalty is measured in dB at the 10^{-12} level. Because the receiver sensitivity operates within parameter limits calculated at an expected 10^{-12} bit error rate, these limits should not be violated by the variability of path parameters. Therefore, attenuation, dispersion, differential group delay, reflections, and so on should not exceed their maximum or minimum limits based on bit rate, fiber length, receiver sensitivity, laser output power, modulation, impairments, and so on. Thus, the maximum optical path penalty (MOPP) is also defined as that penalty that starts deteriorating the bit error rate objective at the receiver.

Thus, receiver design is a complex matter and its parameters and certain nominal values have been the subject of definition by standards such as ITU-T G. 693. Some parameters are the receiver sensitivity, overload, path penalty, efficiency, gain, filtering, and noise (shot and dark noise).

3.3.1 Detection

In optical communications, a typical receiver is sensitive enough to detect the amplitude of the incoming binary optical signal and convert it to a meaningful electrical signal. That is, to distinguish the light intensity between a "logic one" symbol and a "logic zero" symbol. The symbol "one" corresponds to a light pulse, and the "zero" to absence of light, as a result of OOK NRZ modulation. In addition, over a long string of symbols the number of "one" and "zero" symbols are equal; thus, the probability of encountering one or the other symbol is equal at any time. To distinguish which symbol has been received, the receiver makes a decision based on a threshold level V_{thr}; if the incoming signal is above the threshold then a "one" is produced by the receiver, and if it is below it, then a "zero," is produced (Figure 3.4).

The key in the above description is to estimate the receiver detection threshold based on certain propagation parameters of the communications medium. In this

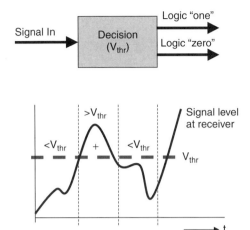

Figure 3.4. A signal at the receiver will be read by a decision circuit as a logic "one" if it is above the threshold level, or "zero" if it is below.

case, one typically invokes probability theory to develop a threshold mathematical expression that can provide a good estimate.

The detection theory in this case starts with two binary variables—the symbol detected is a "one"/"zero," and the symbol transmitted was a "one"/"zero," or a light pulse was "present"/"absent." Based on this, there are four outcomes, two of which are erroneous and two correct:

- symbol detected as "one," whereas transmitted symbol was a "one" (which is correct)
- symbol detected as "one," whereas transmitted symbol was a "zero" (which is wrong)
- symbol detected as "zero," whereas transmitted symbol was a "zero" (which is correct)
- symbol detected as "zero," whereas transmitted symbol was a "one" (which is wrong)

At the detector, let us denote the detected symbol as "one" and as "zero" by D1 and D0, respectively; and let us denote the transmitted symbol as "one" and "zero" by T1 and T0, respectively.

We have made the assumption that over a long string of bits, the number of symbols T_0 and T_1 is equal; that is, both have the same probability of occurrence. This assumption over a long string of symbols is true since the signal is coded and scrambled in order to eliminate long strings of zeros and ones that could potentially disturb the balance of frequency of occurrence of the two symbols. Thus (due to symmetry), the probability of error of symbol T_1 is equal to the probability of error of symbol T_0, and the sum of probabilities for T_1 and T_0 is one:

$$P(T_1) = P(T_0) = 0.5$$

and

$$P(T_1) + P(T_0) = 1$$

Similarly, the sum of all four join probabilities is

$$P(D_1, T_1) + P(D_0, T_0) + P(D_1, T_0) + P(D_0, T_1) = 1$$

Based on the four outcomes listed above, the conditional probability for each of the four cases is $P(D_1|T_1)$, $P(D_1|T_0)$, $P(D_0|T_0)$, and $P(D_0|T_1)$ (Figure 3.5), where for example, $P(D_1|T_0)$ means the probability of detecting a symbol as logic "one" when the symbol transmitted was "zero" (there were no photons in that symbol). Using the conservation of probability, then:

$$P(D_0|T_0) + P(D_1|T_0) = 1 \quad \text{for symbol "zero"}$$

and

$$P(D_1|T_1) + P(D_0|T_1) = 1 \quad \text{for symbol "one"}.$$

If we employ the probability density function $p(x)$ of a random variable x, then the PDF for a sample x given that T_0 is true is $p(x|T_0)$, and that T_1 is true it is $p(x|T_1)$. For Gaussian distributions, the latter two are illustrated in Figure 3.6

Based on this, the probabilities of erroneously detected symbols $P(D_1|T_0)$ and $P(D_0|T_1)$ are the areas in the two tails in the overlapping area of the two distributions (Figure 3.7A). It is obvious that if the distributions are narrower, then the probability for error descreases (Figure 3.7B).

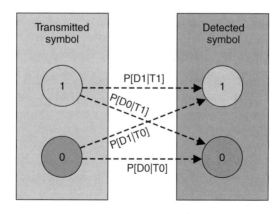

Figure 3.5. Possible outcomes at the receiver.

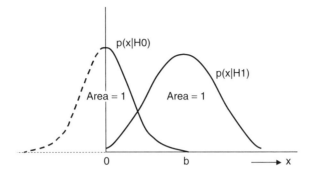

Figure 3.6. Gaussian distributions for peak-pulse detection.

Consider that the conditional probability that the symbol T_1 arrives when a sample x is taken is $P(T_1|x)$, and that the probability that symbol T_0 arrives when a sample is taken is $P(T_0|x)$. A decision as to whether the sampled symbol is detected as "one" or "zero" is based on whether the sample is above or below the threshold level, as already described. Therefore, the decision threshold condition is expressed by the inequality

$$P(T_1|x) \geqq P(T_0|x), \text{ if } H_1, \text{ and}$$

$$P(T_0|x) > P(T_1|x), \text{ if } H_0$$

Consider the property of conditional probabilities $P(x|y)P(y) = P(y|x)P(x)$. Since we deal with a random variable x and its probability density function $p(x)$, $P(x|T_1)$ and

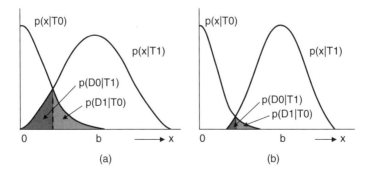

Figure 3.7. Based on the distribution of peak detection, the probability of error or erroneously detected symbols (shaded areas) depend on the tails of the two distributions with respect to the threshold level. In case A, the probability for error is much higher than in case B.

$P(x|T_0)$ can be replaced by their corresponding PDFs (also known as likelihood functions), $p(x|T_1)$ and $p(x|T_0)$, and $P(x)$ by $p(x)$. Then, two ratios are obtained:

$$p(x|T_1)/p(x|T_0) \geq P(T_0)/P(T_1) \qquad \text{if } T_1$$

and

$$p(x|T_1)/p(x|T_0) < P(T_0)/P(T_1) \qquad \text{if } T_0$$

The ratio $L(x) = p(x|T_1)/p(x|T_0)$ is also known as the likehood ratio and it is related to the threshold value $x = V_{thr}$, below which a logic "zero" is detected and above which a logic "one" is detected (Figure 3.8).

Now, if the received signal contains random noise, the Gaussian PDF for symbols "one" and "zero" (and for OOK NRZ) are described by

$$p(x|T_1) = [1/\sqrt{2\pi\sigma_{s+n}^2}] \exp[-(x-b)^2/2\sigma_{s+n}^2]$$

$$p(x|T_0) = (1/\sqrt{2\pi\sigma_n^2}) \exp[-x^2/2\sigma_n^2]$$

where σ_n^2 is the noise variance, σ_{s+n}^2 is the signal-plus-noise variance, and b is the mean signal level. The likelihood ratio is obtained from the ratio of these two expressions. Substituting in the likelihood ratio the values $\sigma_n^2 = 1$, $\sigma_{s+n}^2 = 2$, and $P(T_1) = P(T_0) = 0.5$, and selecting the positive solution from the two possible solutions (when solving the transcendental equation), an inequality $x \geq V_{thr}$ is obtained, where

$$V_{thr} = -(b/3) + (2/3)\sqrt{b^2 + 6 \ln[2P(T_0)/P(T_1)]} = -(b/3) + (2/3)\sqrt{b^2 + 6 \ln 2}$$

When $P(T_1)/P(T_0) = 1$, the area under the PDF $p(x|T_1)$ to the right of the threshold line (Figure 3.8) is equal to the area under the curve $p(x|T_0)$ to the left of it. This condition is known as the ideal observer test.

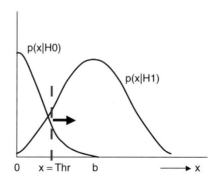

Figure 3.8. Gaussian distributions for pulse detection and threshold level.

In terms of the photonic exposure of the detector, if the average power of the signal at the receiver is P, the period of a bit is T (the inverse of bit rate), and the energy of a photon is $h\nu$, then the probability that n photons will arrive at the receiver is

$$P(n) = [(PT/h\nu)n/n!]e^{-(PT/h\nu)}$$

and the probability of receiving any photons during a "zero" symbol is

$$P(0) = e^{-(PT/h\nu)}$$

Assuming that the probability $P(T_1) = P(T_0)$, then the average probability is $\frac{1}{2}e^{-(PT/h\nu)}$ and this is known as the quantum limit. In optical communications, the receiver sensitivity is typically set at a minimum mean received power so that the incoming signal is reproduced with 10^{-12} bit error rate accuracy over the lifespan of the receiver. Now, in the previous relationships, the ratio $PT/h\nu$ represented the average number of photons for the duration of a bit. It turns out that for highly sensitive receiver photodectors, and at 10^{-12} bit error rate, the minimum number of photons per bit is 57. Thus, the minimum average power per bit is $P = 57\, h\nu/T$.

Similarly, receiver overload is the maximum average received power that is acceptable by the receiver to detect and reproduce the incoming signal at the 10^{-12} bit error rate.

3.3.2 Photodetectors

In general, a p–n diode semiconductor material has one part doped with positive ions (p-type) and another with negative ions (n-type) (Figure 3.9). The contact of the p- and n-types, or the junction, creates a diffusion area. Because of the opposite

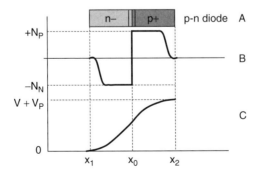

Figure 3.9. (A) The p–n diode semiconductor material has one part doped with positive ions (p-type) and the other with negative ions (n-type). (b) A potential V_d develops across the junction. (C) Applying a bias voltage, V, causes the electron–hole pairs to move, with a mobility that depends on many factors.

polarity of concentrated ions across it, a potential, V_d, develops so that a high concentration of electrons and holes is formed across the junction. From an energy-level point of view, the Fermi level of the p- and a n-types align but the conduction and valence levels of the two do not (the p is higher) and a potential difference between the two is created, known as the *depletion layer*. The energy difference between valence and conduction energy bands is known as the *band gap*, and for different crystals it has different value; for Ge it is 0.67 eV, for Si 1.12 eV, for InP 1.35, and for GaAs 1.42 eV. The two levels are aligned by an externally applied forward bias (voltage), or are further separated by a reverse bias.

In certain p-n junctions, electrons are excited by photons from the valence to the conduction band. Clearly, the energy of the photon must be equal or greater than the energy between the valence and conduction bands, or the *energy gap, Eg*. This defines the minimum energy of photons, or the longest wavelength (also known as *critical* or *cut-off wavelength*), above which neither excitation nor absorption takes place; in such case the junction appears transparent to the photon. Thus, the longest wavelength that establishes the critical wavelength for a particular junction is derived from the well-known relationship $Eg = h\nu = hc/\lambda$, where $hc = 1.24$ eV · μm and λ is the critical wavelength. Since Eg is not the same for all materials (Ge, In, P, As, Ga, etc.), the critical wavelength depends on the type of materials that comprise the p–n junction. There are several compound materials that can absorb light and thus are suitable as photodetectors, such as ZnSe, GaAs, CdS, InP, InAs. However, in fiber communications only those materials that absorb photons of wavelength between 0.8–1.65 μm are usable, and among them are GaAs and InP. As an example, Eg for GaAs is approximately 1.42 eV, and the cut-off wavelength is at 1.24 (eV · μm)/1.42 (eV) = 0.87 μm, which indicates that GaAs is transparent to wavelengths greater than 0.87 μm, and, thus, they cannot be detected. Therefore, a better matching compound is InGaAs.

The key parameters that characterize the photodetectors in communications are the *spectral response, photosensitivity, quantum efficiency, dark current, forward-biased noise, noise-equivalent power, terminal capacitance, timing response* (rise time and fall time), *frequency bandwidth*, and *cut-off frequency*.

- *Spectral response* relates the amount of current produced with the wavelengths of impinging light. Different materials respond differently to electromagnetic radiation. Si detectors respond well to short wavelengths (0.5–1.2 μm), are compatible with integrated Si devices and inexpensive, but they fall short in DWDM applications (1.3–1.6 μm). Similarly, InSb responds to a spectrum about three times as wide as that of Si (0.5–5.1 μm).
- *Photosensitivity* is the ratio of optical power (in watts) incident on the device to the resulting current (in amperes); also known as *responsivity* (measured in A/W).
- *Absolute spectral power responsivity* is the ratio of the output photocurrent (in amperes) from the photodetector to the spectral radiant flux (in watts) at the input of the photodetector.

- *Sensitivity* (in dBm) is the minimum input optical power detected by the receiver (at a 10^{-12} BER).
- At a specific high bit rate, the output photocurrent swing falls to half its maximum. This maximum bandwidth is known as the *3-dB bandwidth*.
- *Quantum efficiency* is the number of generated electron–hole pairs (i.e., current) divided by the number of photons.
- *Dark current* is the amount of current that flows through the photodiode in the absence of any light (hence, dark), when the diode is reverse-biased. This is a source of noise under reversed bias conditions.
- *Forward-biased noise* is a (current) source of noise that is related to the shunt resistance of the device. The shunt resistance is defined as the ratio of voltage (near 0 V) to the amount of current generated. This is also called *shunt resistance* noise.
- *Noise-equivalent power* is defined as the amount of light (at a given wavelength) that is equivalent to the noise level of the device.
- *Timing response* of the photodetector is defined as the time needed for the output signal to rise from 10% to 90% of its amplitude (also known as rise time) and fall from 90% to 10% (also known as fall time).
- *Terminal capacitance* is the capacitance from the p–n junction of the diode to the connectors of the device; it limits the response time of the photodetector.
- *Frequency bandwidth* is defined as the frequency (or wavelength) range in which the photodetector is sensitive. The frequency sensitivity boundaries are found from the wavelength with maximum power level and at a power drop measured in decibels, such as 3 db down, or measured as a percentage, such as 10% down.
- *Cut-off frequency* is the highest frequency (longest wavelength) at which the photodetector is (meaningfully) sensitive.

In summary, as photons enter the biased diffusion area, electron–hole pairs are created, which, being in a very intense electric field, move fast toward the terminals, generating a current. This current is proportional to the number of photons. The responsivity, mobility, and gain of the photodetector depend on many factors. Fast response is related to mobility and speed or recombination of the electron–hole pairs. In communications, photodetectors with fast response and high sensitivity are very critical for multimegabit rates. Such photodetectors are the semiconductor *positive intrinsic negative* (PIN) photodiode and the *avalanche photodiode* (APD).

3.3.3 PIN Photodiodes

The PIN semiconductor photodiode consists of an intrinsic (lightly doped) region sandwiched between a p-type and an n-type. When it is reversed biased, its internal impedance is almost infinite (such as an open circuit) and its output current is proportional to the input optical power.

The input–output relationships that define the *responsivity, R*, and the *quantum efficiency, η*, of the photodiode are

$$R = (\text{output current } I)/(\text{input optical power } P) \quad (\text{amperes/watts})$$

and

$$\eta = (\text{number of output electrons})/(\text{Number of input photons})$$

The quantities R and η are related through the relationship

$$R = (e\eta)/(h\nu)$$

where, e is the electron charge, h is Planck's constant, and ν is the light frequency.

When a photon creates an electron–hole pair, the PIN produces a current pulse with duration and shape that depends on the R × C (resistance times capacitance) time constant of the PIN device and charge-drift diffusion time. The capacitance of the reversed-biased PIN photodiode is a limiting factor to its response (and switching speed). At low bit rates (<Gb/s) and, thus, switching speeds, the parasitic inductance of the PIN may be neglected. However, as the bit rate becomes very high (well exceeding Gb/s), parasitic inductance becomes significant.

3.3.4 APD Photodiodes

The avalanche photodiode (APD) is a semiconductor device equivalent to a photomultiplier. It consists of a two-layer semiconductor sandwich where the upper layer is n-doped and the lower one heavily p-doped. At the junction, charge migration (electrons from the *n* and holes from the p) creates a depletion region, and from the distribution of charges a field is created in the direction of the p-layer. When reverse biased is applied and no light impinges on the device, then, due to thermal generation of electrons, a current is produced, known as "dark current," which is manifested as noise. If the reverse-biased device is exposed to light, then photons cause electron–hole pairs. However, because of the strong field in the APD junction, the pair flows through the junction in an accelerated mode. In fact, electrons gain enough energy to cause secondary electron–hole pairs, which, in turn, cause more. Thus, a multiplication or *avalanche* process takes place. Now, the generated electrons build up a charge, and if the bias voltage is below the breakdown point, the built-up charge creates a potential that counteracts the avalanche mechanism and, thus, the avalanche ceases. If the bias is above the breakdown voltage, then the avalanche process continues and from a single photon a large current is obtained.

There are three types of APD structures:

1. The *deep-diffusion* type has a deep n-layer, compared with the p-layer, and such a resistivity that the breakdown voltage is at about 2 kV. Thus, a wide

depletion layer is created, and more electrons than holes, reducing the dark current. In general, this type has high gain at wavelengths shorter than 900 nm and switching speed no faster than 10 ns. At longer wavelengths, the speed increases but the gain decreases.

2. The *reach-through* type has a narrow junction and, thus, photons travel a very short distance until they are absorbed by the p-type to generate electron–hole pairs. These devices have uniform gain, low noise, and fast response.

3. The *superionization* type is similar to the reach-through type, with a structure that cause the accelerating field to gradually increase. As a result, a low-ionization ratio of holes to electrons (for a certain amount of incident light) is achieved, increasing gain, carrier mobility (electrons are faster than holes), and switching speed.

During this multiplication (avalanche) process, shot noise N_{SHOT} is also multiplied and is estimated as

$$N_{SHOT} = 2eIG^2F$$

where F is the APD noise factor and G is the APD gain expressed as

$$G = I_{APD}/I_{primary}$$

I_{APD} is the APD output current and $I_{primary}$ is the current due to photon–electron conversion.

If τ is the effective transit time through the avalanche region, the APD bandwidth is approximated as

$$B_{APD} = 1/(2\pi G\tau)$$

APDs may be made with silicon, with germanium, or indium–gallium arsenide. However, all types do not have the same performance and characteristics. The type of material determines, among other things, the responsivity, gain, noise characteristics, and switching speed of the device. Thus, although indium–gallium arsenide responds well in the range from 900 to 1700 nm, has low noise, fast switching speeds, and is relatively expensive, silicon responds in the range from 400 to 1100 nm, is very inexpensive, and is easily integrable with other silicon devices; germanium is a compromise between these two.

3.3.5 Photodetector Figure of Merit

The following is a summary of three figures of merit (FOM) which are important in the performance evaluation of photodetectors. Additional specific figures of merit may also be provided by photodetector manufacturers.

- Responsivity $(R) = V_s/(HA_d)$ (VW^{-1})
- Noise-equivalent power (NEP) = $HA_d(V_n/V_s)$ (W)
- Detectivity $(D) = 1/$(NEP)   (W^{-1})

Where V_s is the root mean square (rms) signal voltage (V), V_n is the rms noise voltage (V), A_d is the detector area (cm^2), and H is irradiance (Wcm^{-2}).

In general, APD photodetectors have a much higher gain than PIN photodetectors, but PINs have a much faster switching speed and, thus, they have been largely deployed in high-bit-rate (40Gb/s) detection, particularly in the waveguide PIN detectors. However, APDs keep improving and it is possible to combine high gain with fast switching speeds.

3.4 DETECTION TECHNIQUES

Optical decoding or demodulation is the complementary function of modulation and it entails detecting the received signal and retrieving the binary coded information from it, based on one of three decoding techniques—OOK, PSK, or FSK (see Chapter 1). There are two detection techniques—coherent and intensity modulation with direct detection (IM/DD).

IM/DD directly detects the photonic intensity that impinges on the photodetector. Thus, as the APD or PIN photodiode is illuminated by photons at the junction, the energy of the photons free electrons from the diffusion area (junction). The electrons move under the bias across the junction with a mobility that depends on many factors (a comprehensive description may be found in many textbooks on semiconductor devices). The number of electrons per photon depends on the quantum efficiency of the photodiode material. The speed of electron generation depends on the structure of the photodiode junction and bias. The gain of the photodiode depends on the photodiode material, the junction structure, and bias.

Coherent heterodyne and *homodyne* detection is a technique that was initially developed for radio communications. In optical transmission, the term "coherent" indicates that the local oscillator at the receiver is a light source with the same optical frequency as the incoming signal. Optical coherent methods improve receiver sensitivity by ~20 dB, allowing longer fibers to be used (by an additional 100 km at 1.55 μm). In addition, with IM/DD, the required channel spacing is in the order of 25 to 100 GHz; with coherent techniques it can be as small as 1 to 10 . A metric of good line coders and transmission media (for 10 Gbit/s) is an acceptable eye diagram at the receiver (see Chapter 7); that is, the uncertainty of state (1 or 0) of the received bits is less than one (1) bit per second per hertz (< 1 bit/s-Hz). Optical communication systems are designed with error rates less than 10^{-12} BER.

However, coherent detection has certain stringent requirements. The spectral width of the local oscillator must be narrow, and the local oscillator must be stable and have low noise characteristics.

3.4.1 Demodulation of the Optical Signal

3.4.1.1 OOK RZ or NRZ. On–off keying demodulators use receivers that directly detect incident photons. The number of incident photons in the time domain generates an electrical signal with similar amplitude fluctuation, with the addition of some electrical noise by the photodetector. When the optical signal has been converted to an electrical signal, high-frequency fluctuation (and noise) is low-pass filtered. The signal is sampled at the rate of the expected incoming bit rate and a threshold to minimize jitter and signal-level uncertainty. However, there are instances when the incident amplitude is ambiguous due to excessive noise and jitter and an erroneous 1 or 0 may be produced.

It should be noticed that a OOK NRZ signal provides photons for the full duration of the bit period, whereas a RZ signal does so for a percent of the period. Popular percentages are 33%, 40%, and 50% (see Figure 1.12). The NRZ or RZ modulation is particularly important in ultrahigh bit rates such as 10 or 40 Gbit/s. For example, a 50% OOK 40 Gbit/s signal has logic "1" illuminated for 12.5 ps, whereas a 33% for 8.25 ps. Such slow reduction produces many fewer photons per bit, which should be matched with the quantum efficiency of the photodetector.

3.4.1.2 PSK and FSK. PSK and FSK demodulation is based on coherent detection. That is, in addition to the received optical signal from the fiber, one or two local (optical) oscillators are required to interferometrically interact with the received optical signal and convert it to OOK.

The received PSK-modulated optical signal, ω_S, is mixed coherently with a locally generated laser light, ω_{LO}, and since both are of the same frequency, they interact interferometrically (see Figure 1.10), so that when both frequencies are in phase, there is a constructive contribution, and when they are not, a destructive one and, thus (ideally), an on–off keying signal is generated. Since the accuracy of this

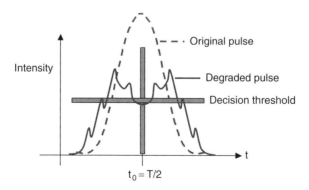

Figure 3.10. A highly distorted pulse may be sampled at an instant at which it produces the wrong symbol. A logic "one," although sampled at the 50% period point, may produce a logic "zero."

method depends on the phase variation of the signal, phase stability and low noise are very critical.

A simpleified homodyne FSK demodulator (see Figure 1.11) passes the FSK-modulated optical signal, ω_S, through a narrow-band optical filter that is tuned to frequency $\omega_1 = \omega + \Delta\omega$. Thus, whenever only this frequency passes the filter, the frequency ω_2 is rejected and the outcome is equivalent to an OOK-modulated signal. Since the accuracy of this method depends on the frequency variation of the signal, no frequency shift (high-frequency stability) and low optical noise are very critical.

3.4.2 Pulse Shape

We said that the incoming signal must be sampled when it is at its highest or lowest expected level. However, there is an issue with this, particularly if the noise is expected to be distorted by noise and other impairments. In this case, the signal may not be exactly at its highest or lowest but in the middle of the pulse (Figure 3.10), in which case the detection may provide the wrong symbol realization; that is, there may be bits in error.

3.4.3 Sampling

Threshold itself, however, is not sufficient to make a clear decision among binary signals. Timing is of the essence since the incoming signal is a string of binary symbols at a fixed bit rate. Thus, the decision is not made only on the basis of threshold level but also at a particular instant of time that has been determined by the timing circuitry of the input port of a node. The timing instant is periodic and it is very important that it be at an instant within the period of the symbol when it is at its highest amplitude (for a logic one) and at its lowest (for a logic zero). However, the in-

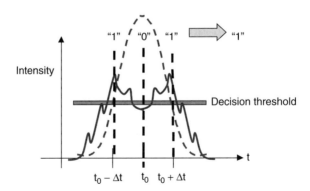

Figure 3.11. Oversampling and a weighting algorithm such as two out of three produces the correct logic value (one or zero).

coming string of symbols is far from ideal and, therefore, the sampling of the incoming signal is not trivial to determine and maintain.

For example, the periodic sampling instants must

- Be exactly at the same frequency as the incoming signal
- Track the phase and frequency variation of the incoming signal
- Maintain its own stability when the incoming signal is stable
- Have the same accuracy as that of the incoming signal

3.4.4 Sampling Methods

Based on this, and depending on the bit rate, different sampling methods have been developed. The simplest of them sets a single sampling point in the middle of the expected pulse, or at 55–60% of the period (past the midpoint). However, this method demands a highly stable clock with critical properties that will be further elaborated on in Chapter 6.

Another method is oversampling. According to this, the expected symbol is sampled at more than one instance within the period. Then, a weighting algorithm such as two out of three is adopted to decide on the symbol logic value (one or zero) (Figure 3.11). However, in this case, the clock must be able to run at several times the incoming frequency, which at ultrahigh speeds presents a real challenge. Similarly, if the modulation scheme is return to zero (RZ), oversampling of ultrahigh bit rates may be technologically impossible.

REFERENCES

1. A. Papoulis, *Probability, Random Variables, and Stochastic Processes,* McGraw-Hill, 1965.
2. C. Ash, *The Probability Tutoring Book,* IEEE Press, 1993.
3. M. Schwartz, *Information Transmission, Modulation, and Noise,* 3rd ed., McGraw-Hill, 1980.
4. R.N. McDonough and A.D. Whalen, *Detection of Signals in Noise,* 2nd ed., Academic Press, 1995.
5. A. Wald, *Statistical Decision Functions,* Wiley, 1950.
6. H. Kogelnik, "On the Propagation of Gaussian Beams of Light Through Lenslike Media Including Those with a Loss or Gain Variation," *Applied Optics, 4,* 12, October 1965, p. 1562.
7. L.W. Casperson and M.S. Sheklami, "Mode properties of annular gain lasers," *Applied Optics, 14,* 11, 2653–2661, November 1975.
8. T.F. Johnson, "Beam Propagation (M^2) Measurement Made Easy As it Gets: The Four-Cuts Method," *Applied Optics, 37,* 21, 4840, July 1998.
9. D.A. Holmes, P.V. Avizonis, and K.H. Wrolstad, "On-Axis Irradiance of a Focused, Apertured Gaussian Beam," *Applied Optics, 9,* 9, 2179–2180, September 1970.

10. J.Y. Wang, "Detection Efficiency of Coherent Optical Radar," *Applied Optics, 23,* 19, 3421, October 1989.
11. H. Nyquist, "Thermal Agitation of Electric Charge in Conductors," *Physical Review, 32,* 10, July 1928.
12. S.O. Rice, "Mathematical Analysis of Random Noise," *Bell System Technology Journal, 23,* 282, 1944.
13. P.P. Webb, R.J. McIntyre, and J. Conradi, "Properties of Avalanche Photodiodes," *RCA Review, 35,* 234–277, June 1974.
14. ITU-T Recommendation G.693, "Optical Interfaces for Intra-office-systems," 11/2001.
15. ITU-T G.707/Y.1322 (2000), "Network Node Interface for the Synchronous Digital Hierarchy (SDH)."
16. ITU-T G.957 (1999), "Optical Interfaces for Equipment and Systems Relating to the Synchronous Digital Hierarchy."
17. ITU-T G.671 (2001), "Transmission Characteristics of Optical Components and Subsystems."

CHAPTER 4

Overview of DWDM Devices and Networks

4.1 INTRODUCTION

In previous chapters, we examined the nature of light and how it interacts with matter as it travels throught it. We also examined how light can be generated with laser devices, how it can be detected, and how it is transmitted through an optically transparent medium, the fiber. We also mentioned in passing a technology that we called dense wavelength division multiplexing (DWDM). With DWDM technology, many optical channels, each at a different wavelength and separately modulated, can be multiplexed and transmitted in the same fiber, thus increasing the aggregate transported bandwidth to astonishing levels that exceed terabits per second. DWDM communications systems require specialized optical and photonic components that are based on light–matter and light–matter–light interactions as well as on the propagation properties of light. These components provide an equivalent functionality of their electrical/electronic counterparts, such as transmitters, receivers, filters, modulators, amplifiers, add/drop multiplexers, cross-connect devices, couplers and so on. Thus, with many optical channels in the same fiber and new components, new phenomena emanate that further degrade the signals due to photon–matter–photon interaction. In this chapter, we review the components that enable this technology, the ramifications of having many optical channels (or DWDM) in the same fiber, and how DWDM technology is employed in modern optical communications networks.

At the outset of this chapter we need to define certain parameters relevant to all optical components and which will be mentioned repeatedly.

Insertion Loss (IL) is defined as the optical power lost due to the intervention of an optical component, or as the ratio of power in to power out. IL is measured in dB, and its mathematical definition is $-10 \log (P_{in}/P_{out})$, where the photonic power is in milliwatts. Insertion loss reduces the optical signal amplitude and it is additive. Over a complete path, many components, connectors, and splices remove some of the signal power due to IL, which contributes to the overall power loss. If the total power loss exceeds the budgeted power, then amplification is required. Thus, the desirable value of insertion loss is near zero. In communications, lower insertion

102 OVERVIEW OF DWDM DEVICES AND NETWORKS

loss allows longer transmission distance before signal amplification and thus fewer amplifiers and reshapers, and, consequently, less maintenance and lower cost.

Isolation is defined as the degree of transmitted power through a component compared with the returned power. The desirable value of isolation is near infinity (or zero power returned through the component). In communications, higher isolation reduces interference as well as potential second-order resonance.

Component isolation should not be confused with *optical channel isolation*. *Optical channel isolation* refers to the spectral gap between neighboring optical channels. In practice, the center wavelengths of DWDM channels are spectrally separated to provide the required isolation. Thus, although *optical channel separation* initially is equal to optical channel isolation, due to dispersion and frequency shifts, the required isolation may be diminished or violated.

The most common parameters of all optical components are

- Insertion loss (IL)
- Optical reflectance
- Operating wavelength range
- Polarization-dependent loss (PDL)
- Polarization-dependant wavelength (variable) (PDW)
- Polarization-dependent reflectance
- Isotropy/anisotropy of refracting material
- Isotropy/anisotropy of diffracting material
- Birefringence of optical material
- Effects of temperature, pressure, and humidity on components
- Effects of vibration (G_{max}) on components
- Effects of electromagnetic fields on components
- Typical and min–max operating conditions
- Mechanical (flex, twist, side, and longitudinal pull, etc.)

The inpact of many of these parameters on the quality of the optical signal is exemplified by problems simulated using the CD-ROM that accompanies this book (see Appendix B for a description of these exercises).

4.2 REVIEW OF DWDM COMPONENTS

4.2.1 The Fiber in Optical DWDM Transmission

Optical fiber is one of the most important parts of the communications network as it impacts the quality of DWDM signals as well as network cost and maintenance. However, even though every care has been taken to manufacture the "perfect" fiber, there are residual imperfections. For example, the physical dimensions of fiber are not exact over its length, although they are contained within the ITU-T defined limits, and the material properties are not uniform, such as the refractive index and index profile. These parameters also vary after the fiber has been manufactured. For example, fiber is not absolutely stress-free (stress causes birefringence) and temperature may affect the propagation properties of fiber, particularly at ultrahigh bit rates. As a consequence,

- The diameter of the core and the cladding varies
- The refractive profile distribution over length varies
- The radial refractive distribution (refractive-index profile) varies
- The density of contaminants or dopants over length varies
- Birefringence varies
- Polarization varies
- Dielectric constant varies with temperature

These imperfections affect the quality of the signal and they are either manifested as power loss, polarization state variation, noise, and other linear and nonlinear effects. In general, good fiber preserves characteristic uniformity for every kilometer of its length. However, fiber technology keeps improving to meet increasing bandwidth needs and spans in optical networking. For example, when the first SMF, optimized for 1310 nm, made its debute in 1983, it supported transmission at 2.5 Gbit/s for up to 640 km without amplification (or 10 Gbit/s up to ~100 km). Recent SMF transports the same optical signal (2.5 Gbit/s) to over 4400 km (or 10 Gbit/s up to ~500 km) without amplification. Similarly, the bandwidths supported by standard SMF was about 1300 and from 1520 to 1620 nm but not about 1400 nm. Currently, SMF fiber supports the compelte spectrum from 1300 to 1620 nm, allowing for more DWDM channels. Thus, the aggregate DWDM bandwidth–span product is calculated by multiplying the bandwidth–span product by the number of optical channels; for 10 Gbit/s, 2000 km, and 160 channels, this is projected to 3,200,000!

Unlike copper wires that can be connected very easily, bringing two fiber ends in contact represents a discontinuity and requires special treatement. For example, the two fibers may be of different types and different refractive index profiles; each fiber may have been supplied by a different manufacturer or at a different time and using different technology (this may impact purity and refractive-index profile). Moreover, the glassy end faces may be flat or not, parallel or not, of different optical quality (with imperfections and contamination on the faces), and between the end faces there may be a minute but finite air gap. Thus, photons traveling from one fiber to another have to overcome these discontinuities and imperfections with minimal loss. Consequently, special precautions must be taken to minimize optical power loss through fiber:

- The two fiber ends to be connected should be treated so that the end faces are flat, perpendicular to fiber's longitudinal axis, and highly polished. Alternatively, the faces may be formed into spherical lenses to concentrate light, but this is a more specialized operation that requires specialized personnel and equipment. The concentricity error of single-mode fiber (based on ITU-T G.652) should be less than 1 μm.

- The two end faces should be brought in close (a small fraction of λ) proximity.
- The two end faces should be treated with antireflecting coatings. Reflection may be deleterious to the laser source and if the gap has the right length, it may act as a filter, attenuating specific wavelengths.
- The two end faces should be free from contaminants. Contaminants may attenuate optical power through the connection.
- The two fiber cores should be in perfect alignment; typically, the cores are precisely aligned with biconical, self-aligned connectors or with aligned groves.
- In some cases, a refractive index-matching fluid may be used.

Another fiber interconnection method splices two fibers by fusion. This operation requires specialized splicing equipment with which the two fiber ends are first stripped from their cladding and then the two cores are fused together and retreated to yield a continuous fiber.

Splicing is preferred when fiber connectivity is permanent (i.e., repairing broken cables) and in the field (aerial or buried cable). Connectors are preferred in central offices, cabinets, and huts where connectors are protected and where disconnecting and reconnecting may be expected.

4.2.2 Optical Multiplexers and Demultiplexers

Optical demultiplexers receive a multiwavelength beam and separate it into its wavelength components. There are *passive* and *active* demultiplexers. Passive demultiplexers are based on prisms, gratings, and spectral (frequency) filters. Active demultiplexers are based on a combination of passive components and tunable detectors, each tuned to detect a specific frequency.

Multiplexers perform the reverse functionality of demultiplexers; they receive several separated wavelengths and form a single beam that consists of all these wavelengths.

Key parameters specific to multiplexers and demultiplexers are

- Wavelength-dependent attenuation
- Far-end crosstalk
- Near-end crosstalk

4.2.3 Optical Filters

Optical filters allow the passage of a narrow band of optical frequencies. In DWDM, a bandpass filter should be as accurate as one part in 10,000 (1$%_{000}$) in wavelength. Optical spectral filters are based on *interference, diffraction,* or *absorption* and they are distinguished as *fixed* and *tunable*.

When the electric field of light interacts with the atoms or molecules of matter, *absorption* takes place. Thus, absorption is wavelength dependent. The amount of light absorbed is proportional to the intensity of incident light. If α is the absorption coefficient, c is the concentration of absorbing centers such as ions (for solid matter $c = 1$), and x is the path traveled in matter, then *transmittance*, T_x, is defined as

$$T_x = 10^{-\alpha c x}$$

The total transmittance, T_T, of a stack of filters is the product of each filter's transmittance. Thus,

$$T_T = T_1 \times T_2 \times \ldots \times T_N$$

The inverse of transmittance is known as *opacity*. Optical density, D, is the logarithm of opacity:

$$D = \log(1/T)$$

The most common parameters specific to filters are

- Backward loss
- Polarization-mode dispersion
- Modulation depth (modulated light)
- Output light intensity (optical power per OCh)
- Finesse (FP filter)
- Spectral width
- Line width
- Cutoff λ
- Extinction ratio
- Line spacing

Spectral width is defined as the band of frequencies that a filter can pass through. The spectral width is characterized by an upper and a lower frequency (wavelength) threshold.

Line width or *channel width* is defined as the width of the frequency channel. An ideal channel would be monochromatic. However, this is not possible and, thus, the line width is a measure of how close to ideal a channel is, as well as an indication of the spectral content of the channel.

Line spacing is defined as the distance in wavelength units (nm) or in frequency units (GHz) between two channels.

106 OVERVIEW OF DWDM DEVICES AND NETWORKS

Fixed optical filters respond to a specific narrow slice of the spectrum (hence, fixed), in contrast to *tunable optical filters* (TOF) that respond to a narrow but selectable slice of a wide spectrum. Tunability is achieved by applying voltage, temperature, heat, stress, pressure, or some other controlling mechanism. Thus, in addition to the aforementioned, TOF-specific parameters are

- Wide tuning range
- Narrow bandwidth
- Fast tuning
- Flat gain
- Low insertion loss
- Low polarization-dependent loss
- Low crosstalk
- Insensitivity to temperature

Certain common parameters pertaining to filters, such as center and peak wavelength, insertion loss, passband, 20 dB bandwidth, and others are shown in Figure 4.1. The bandwidth difference ΔBW = (20 dB bandwidth) – (1 dB bandwidth) is a measure of *steepness* and an indication of rejection of signals outside the filter passband; the smaller the difference the better. In addition, *slope of the filter* in $\Delta \lambda$ is calculated from the wavelengths corresponding to 80% of the peak transmission and the 5% absolute transmission; the lower and higher (or left and right) slopes are not necessarily symmetric but yield two different slope measurements. The slope is expressed in percent and is calculated by

$$\text{slope} = [\{\lambda \,(80\% \text{ of peak}) - \lambda \,(5\% \text{ absolute})\}/\lambda \,(5\% \text{ absolute})] \times 100 \,(\%)$$

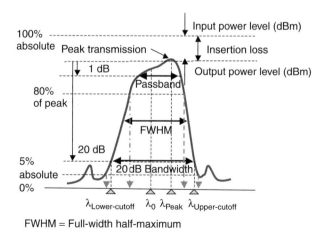

Figure 4.1. Filter-parameter definition.

Among the different types of filters, the most popular are

- The Fabry–Perot
- The dielectric thin film
- The diffraction Bragg grating and the fiber Bragg grating
- The arrayed waveguide grating
- The Mach–Zehnder, fixed and tunable
- The acoustooptic tunable

A full description of these is beyond the purpose of this book but can be found in several other books on DWDM technology.

4.2.4 Rotators—The Faraday Effect

Certain materials exhibit a particularly optical nonlinear behavior manifested as rotation of the polarization state of light. This phenomenon is known as the *Faraday effect* or *photorefraction*. Devices based on this are known as *rotators*. The rotation of polarization state is also known as *polarization mode shift*, θ, and is explained as follows.

Plane-polarized light, upon entrance into the *photorefractive* medium, is decomposed into two circularly polarized waves rotating in opposite directions and traveling in the medium at different speeds. Thus, a phase between the two is created, increasing linearly the more it travels in the medium. The two waves recombine upon exiting the medium, producing a wave but with a polarization plane shifted (or rotated) by an angle relative to its polarization at the entrance. The amount of mode shift depends on the thickness of the medium, d (cm), on the magnetic field H (oersted), and on a material constant V, known as the *Verdet constant* (min/cm-Oe), expressed by

$$\theta = VHd$$

4.2.5 Optical Power Attenuators

Optical power attenuation is an impediment in optical transmission. Despite this, components with controlled attenuation, known as *optical power attenuators* (OPA), play an important role in the design of optical communication systems. They are used to adjust the optical power of selected DWDM channels at the input and/or output of optical amplifiers so that the power differential among DWDM channels is minimal and their power level matches the dynamic range of the receiver.

Optical attenuators may be fixed or variable; *variable* optical attenuators (VOA) are controlled either mechanically or electrically.

The desired characteristics of variable attenuators are

- Small size
- Polarization insensitivity

- Dynamic power attenuation range (1–40 dB)
- Wide operating wavelength range (1200–1600 nm)
- Resolution (0.1–0.2 dB) or incremental attenuation
- Low optical return loss (>50 dB)
- Accuracy
- Tolerance to insertion loss
- Repeatability
- Stability
- wide operating-temperature range (0–65°C)
- Low cost

4.2.6 Optical Isolators and Circulators

Optical isolators are two-port devices that allow optical power (of a spectral band) to flow unidirectional from one terminal to the other, a function similar to the electrical diode.

Optical isolators are characterized by

- *Insertion loss, L*, or the loss of optical power through it
- *Isolation, I*, or the ratio of transmitted power in the desired direction over the other

The quantities L and I are expressed by

$$L(\text{dB}) = P_I(\text{dB}) - P_T(\text{dB}),$$

and

$$I(\text{dB}) = P_I(\text{dB}) - P_R(\text{dB})$$

where P_I, P_T, and P_R are the incident power, the transmitted power and the reflected power, respectively, all expressed in decibel units.

More than one isolator may be connected to form a three-terminal device, permitting unidirectional energy flow from terminal 1 to 2, from 2 to 3, and from 3 to 1. This device is known as a *circulator*. When a signal is transmitted from port 1 to port 2, another signal is not restricted from passing from port 2 to port 3, or 3 to 1.

Among the desirable characteristics of circulators are

- Small size
- High isolation (>50 dB)
- Low insertion loss (<1dB)
- Low return loss (>50 dB)
- Directivity ($\ll -50$dB)

- Low polarization sensitivity
- Low PDL loss (<0.05 dB)
- Low PMD (<0.1 ps)
- High optical power (>300 mW)
- Environmental stability over the specified wavelength range
- Wide operating temperature range (–20°C to + 60°C).

4.2.7 Gratings

The principles of diffraction are used in *diffraction gratings*. A diffraction grating can be used as a filter, as a multiplexer/demultiplexer, and as a switching device. In a simple implementation, a glass substrate with adjacent epoxy strips is blazed to form a (planar) *diffraction grating*. There are two fundamental planar grating types: the *reflection* grating and the *pass-through* grating. A diffraction grating works as follows. Incident rays interact with the edges of the groove and they are either reflected or pass through (depending on the diffraction grating type). The reflected rays have a path difference that represents a delay or a phase difference, equal to $m\lambda$, where m is an integer and λ the wavelength. These rays recombine and, depending on distance, geometry of the grooves, and angle of incidence, undergo constructive or destructive interference according to the wavelength and delay of reflected rays by adjacent grooves. Thus, each wavelength component is diffracted at different angles, according to

$$d(\sin \alpha + \sin \beta) = m\lambda$$

where α is the angle of incidence, β the angle of diffraction, and m is the diffraction order (or spectral order) that takes positive and negative integer values $0, \pm 1, \pm 2, \ldots$. Thus, when a polychromatic light beam impinges on a diffraction grating, each wavelength component is diffracted at selected angles. It is this spatial distribution of wavelengths that make the diffraction grating useful in a variety of applications.

When $m = 0$, the diffraction grating acts like a mirror to the incident beam; that is, wavelengths are not separated. This is known as "*zero-order diffraction*" or "*specular reflection*" and the grating acts like a mirror.

The nonzero values of m define the diffracted angular direction of wavelengths. By definition, positive direction is the direction for which m has positive values, $\beta > -\alpha$, and with respect to the normal axis of the grating plane, the wavelengths are reflected at the same side of the normal axis (Figure 4.2). Similarly, negative direction is the direction for which m has negative values, $\beta < -\alpha$, and with respect to the normal axis of the grating, the wavelengths are reflected at the other side of the normal axis.

Gratings diffract a wavelength with both positive and negative orders, for which

$$-2d < m\lambda < 2d$$

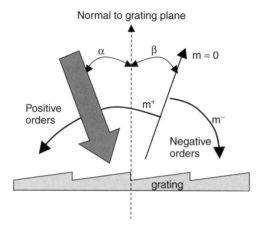

Figure 4.2. Definition of postive and negative diffraction modes.

When $\lambda/d \ll 1$, a large number of diffracted orders exist. Typically, few orders are desirable and, thus, the grating spacing should be comparable to the wavelength. For realizable gratings, the empirical relationship is useful:

$$|m\lambda/d| < 2$$

The latter relationship, based on wavelength and grating spacing, limits the number of values or *orders* of the integer m.

Notice that in the grating relationship $d(\sin \alpha + \sin \beta) = m\lambda$, the parameter d is a constant. Thus, for a fixed angle of incidence α, the grating relationship is satisfied by several wavelengths. In fact, two or more wavelengths can be diffracted in the same direction, each satisfying the grating relationship for a different value of m. This is an undesirable case of wavelength overlapping. However, it should also be noticed that in practice the angles α and β are smaller than 90°. That is, $\sin \alpha < 1$ and $\sin \beta < 1$, hence $\sin \alpha + \sin \beta = <2$, and, thus, the grating relationship is valid only for those values of m for which $|m\lambda/d| < 2$.

Now, the *angular dispersion, D*, between wavelengths for a given order m is obtained from the diffraction equation by differentiating β for λ. Doing so, one obtains

$$D = d\beta/d\lambda = m/(d \cos \beta)$$

That is, the angular dispersion increases as d decreases.

Similarly, the *linear dispersion* of the grating provides an indication of how far apart in space two wavelengths are in the focal plane, that is,

$$D = dx/d\lambda$$

The inverse of this relationship, known as *spectral resolution, R*, indicates how many wavelengths per unit of distance in the focal plane there are:

$$R = d\lambda/dx$$

The *absolute efficiency* of a grating is defined as the energy flow (or power) of the diffracted monochromatic light of order m relative to the energy flow of the incident light. The efficiency of gratings depends on the

- Type of grating (reflective or transmission)
- Grating material
- Density of grooves
- Uniform periodicity of grooves
- Paralleleity of grooves
- Geometry of grooves
- Type of coating of grooves
- Reflection efficiency of grating
- Conductivity of groove surface
- Flatness of grating (unless concave gratings are examined)
- Angle of incidence
- Apodisation
- Spectral content of the beam
- Order m for which efficiency is measured
- Polarization of light

Gratings are used in a variety of applications, such as

- Optical multiplexers/demultiplexers
- Optical cross-connects (in conjunction with beam deflectors)
- Optical channel add/drop
- Filters
    Narrow bandpass
    Band rejection (including ASE noise suppression)
- Dispersion compensation
- Pulse reshaping (including pulse-broadening compensation)
- Optical taps
- Short-wavelength sensor gratings
- Mode and polarization converters
- Distributed-feedback laser-wavelength stabilizers
- Monochromators, spectrographs, and sensors

4.2.7.1 Fiber Bragg Gratings.

A *fiber Bragg grating* (FBG) consists of a fiber segment whose index of refraction varies periodically along its core length. It behaves as wavelength-selective reflector and finds many applications, as listed in the previous section.

The reflectivity of Bragg gratings for a given mode at center wavelength is given by

$$R = \tanh^2 [\pi \cdot L \cdot \Delta n \cdot \eta(V)/\lambda_b]$$

where L is the length of the grating, Δn is the magnitude of index perturbation, and $\eta(V)$ a function of the fiber parameter V that represents the fraction of the integrated mode intensity contained in the core.

Fiber gratings assume a single mode polarization-maintaining fiber with a circular core. However, when the core is (imperfectly) elliptical, then the grating supports two propagation modes with each polarization axis having a different propagation constant.

4.2.7.2 Dependence of Gratings on Temperature.

A diffraction grating consists of matter, the optical parameters of which are affected by heat. As the grating is heated, matter expands, the grating constant (number of lines per unit length) changes, and so do its diffraction specifications. In this example, we consider the reflection-type grating (the transmitted and the blazed-fiber types may also be considered), as well as the fiber Bragg grating. For clarity, we repeat some of the relationships already provided.

The blaze or Bragg angle Θ_B, the wavelength λ, and the line spacing a are related by

$$\Theta_B = 1/\sin(\lambda/2a)$$

The diffraction angle β is related to the incident angle α and the wavelength by

$$\sin \beta = \sin \alpha \pm (\lambda/n_0 a)$$

where $a = 2\pi/K$, and K is the grating wave number.

For a fixed incident angle α, the variation of λ, $\Delta\lambda$, with respect to the Bragg angle is

$$\beta = \Theta_B + \Delta\lambda/(n_0 a \cos \Theta_B)$$

As the Bragg grating temperature changes, due to the variation of the grating constant a stationary observer sees a wavelength shift by $\Delta\lambda_0$ from the original wavelength λ_0. This change is given by

$$\Delta\lambda_0 = \zeta \Delta T \lambda_0$$

where ζ is the thermal expansion coefficient for the grating material (for silica it is 0.55×10^{-6}). In case of a fiber Bragg grating, the above is modified as:

$$\Delta\lambda_0/\lambda_0 = (\zeta + \xi)\Delta T$$

where ξ is the thermooptic coefficient (~8.3×10^{-6} for germanium-doped silica core) of the fiber Bragg grating.

The effect of the thermal expansion on the Bragg spacing is

$$\Delta a = \zeta \Delta T$$

Thus, for a fixed λ, λ_{fixed}, and for a fixed incident angle α, the variation of β, $\Delta\beta$, with respect to the Bragg spacing a is

$$\sin\beta = \sin\alpha \pm (\lambda/n_0 a)$$

Applying perturbation theory,

$$\Delta\beta/\Delta T \cos\beta = \pm\Delta(\lambda/n_0 a)/\Delta T$$

or

$$\Delta\beta/\Delta T \cos\beta = \pm(\lambda/n_0)\Delta a/\Delta T$$

where the term a is constant and thus is eliminated.

For very small angles of diffraction β, this is simplified to

$$\Delta\beta/\Delta T = \pm\zeta\lambda/n_0$$

That is, the spatial distribution of wavelengths (or the angle of diffraction) depends on the Bragg grating temperature expansion coefficient.

4.2.8 Optical Amplifiers

Optical signal attenuation is compensated by optical amplification so that the signal at the receiver arrives at a power greater than the receiver sensitivity and also at the expected low bit error rate (~10^{-12}).

For instance, if the power at the source is P_S (dBm), the expected power level at the receiver is P_R (dBm), and the fiber attenuation coefficient is α (db/km), then, ignoring other types of losses (dispersion, polarization, etc.), the allowable fiber length is

$$L = (P_S - P_R)/\alpha \text{ (km)}$$

Attenuation puts a limit on fiber length between source and receiver, unless the optical power of the signal is amplified, typically every 40 to 80 km, compensating for losses and thus extending the source–receiver distance.

Recently, direct amplification technologies have been developed and used to directly amplify a weak optical signal. To date, the most commonly used *optical amplifiers* (OA) are the *semiconductor optical amplifiers* (SOA), the *optical fiber amplifiers* (OFA), and the *Raman amplifiers*. The key common characteristics of optical amplifiers are *gain, gain efficiency, gain bandwidth, gain saturation, noise, polarization sensitivity,* and *output saturation power*. Among other characteristics are *sensitivity (gain and spectral response) to temperature* and other environmental conditions, *dynamic range, cross talk, noise figure, stability,* and *physical size*.

- *Gain* is the ratio of output power to input power (measured in dB).
- *Gain efficiency* is the gain as a function of input power (dB/mW).
- *Bandwidth* is a function of frequency and, as such, *gain bandwidth* is the range of frequencies over which the amplifier is effective.
- *Gain saturation* is the maximum output power of the amplifier, beyond which it cannot increase despite increase in the input power
- *Noise* is an inherent characteristic of amplifiers. In electronic amplifiers, noise is due to (random) spontaneous recombination of electron–hole pairs that produces an undesired signal added to the information signal to be amplified. In optical amplifiers, it is due to spontaneous light emission of excited ions, which we will further explore.
- *Polarization sensitivity* is the gain dependence of optical amplifiers on the polarization of the signal.
- *Output saturation power* is defined as the output power level for which the amplifier gain has dropped by 3 dB.

Optical amplifiers introduce noise that contaminates the optical signal and thus decreases the signal-to-noise ratio (SNR). Therefore, it is important that the noise characteristics of amplifiers are specified. The noise figure (NF) is the dimensionless ratio of input SNR to output SNR of an amplifier:

$$NF = SNR_{in}/SNR_{out}$$

4.2.8.1 Semiconductor Optical Amplifiers. Semiconductor optical amplifiers (SOAs) are based on conventional laser principles; an active waveguide region is sandwiched between a p- and n-region. A bias voltage excites ions in the region, which create excited electron–hole pairs. Then, as light is coupled in the active waveguide, it stimulates the excited electron–hole pairs and they recombine and generate more photons (of the same wavelength as the optical signal) and, hence, optical amplification is achieved. In a different implementation, the active region is illuminated with photonic energy to cause the necessary excitation in the active region. The amplifier gain, G, of the amplifier typically exhibits a wavelength dependence approximated to

$$G = -12[(\lambda - \lambda_p)/\Delta\lambda]^2 + G_p$$

where G_p is the peak gain at the corresponding wavelength λ_p, $\Delta\lambda$ is the full-width, half-maximum (FWHM) gain bandwidth, and the factor 12 is a result of the definition of $\Delta\lambda$.

The 3 dB saturation output power, P_s, is described as a function of λ by

$$P_s = q_s(\lambda - \lambda_p) + P_{s-p}$$

where q_s is a linear coefficient and P_{s-p} is the 3 dB saturation output at the peak gain wavelength λ_p.

4.2.8.2 Optical Fiber Amplifiers. Optical fiber amplifiers (OFAs) are heavily doped fibers with a rare-earth element such as erbium. Erbium atoms are excited by high-energy light (980 and/or 1480 nm) and they emit photons in the 1530 to 1565 nm range. Quantum yield, Φ, is a measure of the efficiency of a photon at a given wavelength for a given reaction and for a given period of time, and it is defined as

$$\Phi = \text{(number of emitted photons)/(number of absorbed photons)}$$

However, the number of absorbed photons, I_a, is not directly proportional to the optical density (OD) at the excitation wavelength but is connected with the relationship

$$I_a = k(1 - 10^{-OD})$$

where k is a constant proportional to incident intensity.

Optical fiber amplifiers (OFA) are classified as *power amplifiers, preamplifiers,* and *line amplifiers*. An optical power amplifier acts like a booster and it is placed right after the source (or the modulator) and thus may be integrated with it. A preamplifier is placed directly before the detector and may be integrated with it. The line amplifier must have high gain and very low noise so that it does not degrade the signal-to-noise ratio of an already attenuated signal. The proper applicability of OFA types is described by ITU-T in G.662 and G.663 so that nonlinearity, polarization, and other effects that affect the integrity of the channel and the quality of the transmitted signal are minimized. In addition, other ITU standards deal with optical amplifiers, such as G.957 (optical-return loss) and G.973 (remotely pumped amplifier). Some of the important parameters in the EDFA gain process are

- Concentration of dopants
- Effective area of EDFA fiber
- Length of EDFA fiber
- Absorption coefficient
- Emission coefficient
- Power of pump
- Power of signal

- Relative population of upper states
- Lifetime at the upper states
- Direction of signal propagation with respect to pump

EDFAs have found applications in WDM long-haul transport systems with gain in excess of 50 dB over a spectral range of 80 nm. However, EDFAs do not exactly amplify all wavelengths the same and, thus, there are three key issues to be addressed as they impact the signal quality, gain flatness, dynamic gain, and low noise.

- EDFA gain is not flat over the spectral range; this is addressed with gain flattening optical filters. In addition, the total EDFA gain is shared by the number of optical channels. Thus, many optical channels share less gain than fewer channels would. This is addressed by using higher-gain EDFA amplifiers. This means that either more powerful pump lasers must be used or a higher concentration of erbium ions. However, as pump power increases, other issues become more critical (laser stabilization, longevity of laser device, and cost of pump). Currently, typical output power of a 980 nm laser pump is about 350 mW and for a 1480 nm pump it is about 250 mW. Increasing the erbium concentration tends to cluster ions, causing the amplifier efficiency to drop and dispersion to increase.
- As the gain of an EDFA is shared by all wavelength channels, the greater the number of channels, the less gain per channel. This has an undesirable effect in optical add–drop multiplexing (OADM); after few optical add–drops in a span, the received wavelengths have different power levels. This is addressed with optical-power equalization.
- EDFA spontaneous emission introduces noise that degrades the S/N ratio. Moreover, optical noise sources are cumulative; the more the EDFAs, the more the noise and the lower the S/N ratio. Thus, one would be inclined to use higher gain to overcome this. However, near the zero-dispersion wavelength region nonlinear effects (such as four-wave mixing) become significant and further degrade the S/N ratio. Thus, the selection of power (per channel) launched into the fiber becomes a puzzle as amplifier noise restricts the minimum power of the signal and four-wave mixing limits the maximum power per channel launched into the fiber.

The aforementioned implies that the power must lie within a lower and an upper limit. To determine the proper power limits, other parameters must be considered, such as

- Fiber length between amplifiers (in kilometers)
- Fiber attenuation (loss) per kilometer
- Number of amplifiers in the optical path
- Amplifier parameters (gain, noise, chromatic dispersion, bandwidth)

- Number of channels (wavelengths) per fiber
- Channel width and channel spacing
- Receiver (detector) specifications
- Transmitter specifications
- Dispersion issues
- Polarization issues
- Nonlinearity issues
- Optical component losses and noise (connectors and other devices)
- Quality of signal (bit error rate, S/N)
- Signal modulation method and bit rate
- Number of amplifiers over the span
- Number of optical add–drop multiplexers
- Many other design parameters

4.2.8.3 Amplified Spontaneous Light Emission. As already mentioned, amplified spontaneous light emission (ASE) in fiber amplification is an issue of concern because it is a serious noise source that limits the number of amplifiers in a span. The generated ASE propagates in both directions in the fiber amplifiers, forward and backward with respect to the pump propagation. Backward ASE is most intense (highest peak) in the first meters, whereas forward ASE is not as intense at the beginning of the fiber amplifier. As forward ASE is generated and propagated, there are two countermechanisms: ASE loss and ASE accumulation. The combination of the two yields a monotonically increasing ASE at the end of few-meters-long fiber. This is where a rejection filter can minimize the transfer of ASE from the fiber amplifier to the transmission fiber.

4.2.8.4 Raman Amplifiers. When two light sources propagate in the same single-mode fiber, a continuous wave of short wavelength and high power (~500 mW) and a signal of longer wavelength and low power are generated. The short-wavelength, high-energy source (known as the pump) excites fiber atoms to a higher energy level which, under the right conditions, is stimulated by the longer-wavelength signal, releasing photons in the same long-wavelength spectrum. The amplification process is based on the nonlinearity of the single-mode fiber and it is evolutionary (that is, there is a gradient gain) (Figure 4.3) and known as *Raman amplification*.

From a classical viewpoint and in terms of signal and pump power, the signal power gradient (from which the Raman gain, G_R, may be obtained) is

$$dP_S(z)/dz = \{[g_0/(KA_{\text{eff}})]P_R(z) - \alpha\}P_S(z)\}$$

where $P_S(z)$ is the signal power at point z, $P_R(z)$ is the power of the Raman pump at point z, g_0 is the Raman gain coefficient, α is the fiber attenuation coefficient, K is a factor (set at the value of 2) and A_{eff} is the fiber-core effective area.

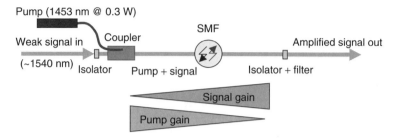

Figure 4.3. Principles of Raman amplification.

Assuming that the Raman gain coefficient is linear, then solving the differential equation (and integrating from 0 to L) one obtains

$$G_R = \exp\{g_0 P_R(L) L_{\text{eff}}/(2A_{\text{eff}})\}$$

Where L_{eff} is the effective length expressed as the integral of $[P_R(z)/P_R(z)]dz$ over the fiber length L. L_{eff} is also given as $[1 - \exp(-\alpha L)]/\alpha$ and it expresses the effective fiber length seen by the Raman gain evolution.

Similarly, if the initial signal power is $P_s(0)$ and the signal power at point z is $P_s(z)$, the Raman power gain is described by the exponential relationship

$$P_s(z) = P_s(0) \exp\{G_R P_p(z) L_{\text{eff}} - \alpha z\}$$

where

- A_{eff} is the effective core area of the pump
- K is a polarization factor that describes how close the polarization state of the pump is to the signal, ranging from orthogonal to parallel
- $G_R = g_R/A_{\text{eff}} K$ is known as the Raman gain efficiency
- g_R is the Raman gain
- $P_p(z)$ is the pump power
- α is the fiber attenuation coefficient
- L_{eff} is the effective fiber length

In DWDM applications, there are many wavelengths within a band, such as C or L, that require amplification. One implication of this is that the Raman effect may cause cross talk if the launched power of each channel exceeds a threshold level. Another implication is that Raman gain is not uniform for all optical channels. In fact, the Raman gain coefficient increases almost linearly with wavelength difference (or offset); it peaks at 100 nm and then rapidly drops (Figure 4.4). As a consequence, the best gain performance is at an offset of 100 nm from the pump wavelength.

Because the usable gain bandwidth is narrow, in the range of 20–40 nm, if a broader amplification band is needed, several "Raman" pumps are used, each offset

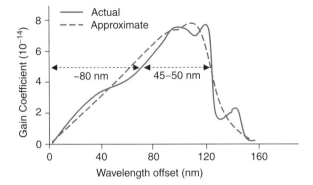

Figure 4.4. Raman gain coefficient versus signal–pump wavelength offset.

by ~40 nm, so that together they can effectively cover the broad spectral band (Figure 4.5). Raman amplification is applicable over a broad spectrum, from 1300 to 1600 + nm, known as the *Raman supercontinuum.* In this spectrum, over 500 optical channels with 100 GHz spacing (or 1000 channels with 50 GHz spacing) may be used, enabling a multi-Terabit DWDM technology.

In a two-pump system, the power and gain of a second pump is estimated from the power and gain of the first pump according to

$$P_{\text{pump2}} = P_{\text{pump1}}(g_2/g_1) \text{ (mW)}$$

Where P_{pump1} and g_1 are the first pump power (in mW) and gain (in dB), P_{pump2} and g_2 are the second pump power (in mW) and gain (in dB).

4.2.8.5 Raman Noise. Because Raman amplification requires long fibers, existing noise in the signal is amplified and more noise is added by the Raman pump,

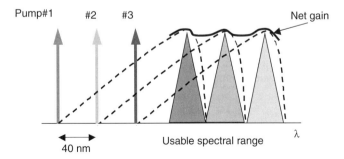

Figure 4.5. Spacing several Raman pumps separated by ~40 nm. Channels in a wider spectral range are amplified.

known as relative intensity noise (RIN). The RIN, R_S, is caused by fluctuations of the Raman pump and it is equal to the ratio of the mean square fluctuation of optical power, $\langle \delta P_S^2 \rangle$, to the square of the mean optical power, $\langle P_S \rangle^2$:

$$R_S = \langle \delta P_S^2 \rangle / \langle P_S \rangle^2$$

Similarly, as the pump induces noise on the signal, the strongly fluctuating (or modulated) signal induces noise on the pump. This relative induced noise, R_P, is expressed by a similar ratio,

$$R_P = \langle \delta P_P^2 \rangle / \langle P_P \rangle^2$$

The RIN, R_S, is expressed in terms of the pump RIN, R_P, and the Raman gain, g_R, dependent term as

$$R_S = R_P + 20 \log [\ln (g_R)] + 10 \log [D, \Delta\lambda, L, \alpha, v_S, v_g, f_P]$$

where the third term, $10 \log [D, \Delta\lambda, L, \alpha, v_S, v_g, f_P]$, is a function that depends on the pump propagation direction (counter- or copropagating), expressed in terms of the fiber length, L, dispersion, D, wavelength difference, $\Delta\lambda$, fiber attenuation constant, a, signal velocity, v_S, group velocity of the pump, v_g, and fluctuation of the pump, f_P.

Some observations may be made now, related to copumped and counterpumped Raman methods:

- The RIN induced on the signal is always greater than the RIN on the pump.
- To maintain the RIN on the pump 30 dB lower than the RIN on the source (assuming that the third term in the above relationship is eliminated), the Raman gain should not be greater than 4.4 dB, that is, $20 \log [\ln (g_R)] = 30$ dB, or $g_R = \sim 4.5$.
- Counterpropagating averages the noise over the transit length of fiber, thus acting as a low-pass filter with an extinction ratio of 20 dB per decade.
- Copropagating averages the noise transfer because of "walk off." In this case, dispersion acts like a low-pass filter with an extinction rate of 20 dB per decade.

Although amplified spontaneous emission (ASE) in optical amplification is unavoidable, Raman ASE is much less than EDFA ASE. The reason is that Raman ASE is a distributed process that in effect takes place over a long length of fiber (tens of kilometers), as opposed to EDFA amplification that takes place over tens of meters. Thus, an equivalent (lumped) noise figure, F_R, is defined as

$$F_R = [(R_{ASE}/hf) + 1]/G_R$$

where R_{ASE} is the Raman ASE density at the end of the fiber, G_R is the Raman on/off gain, h is Planck's constant, and f is the optical frequency (Raman on/off gain is defined as the ratio of signal power when the Raman pump power is on over the signal power when the Raman pump power is off).

In such a case, the ASE spectral density, ρ_{ASE}, is

$$\rho_{ASE} = h\nu(G_R F_R - 1)$$

where h is the Planck's constant and n is the frequency.

Another noise source is the *double Rayleigh scattering* (DRS), which is proportional to fiber length. As a consequence, the contribution of DRS noise is limited by lowering the Raman pump power and, thus, the Raman gain. Therefore, the DRS effectively puts a limit on Raman gain per stage.

The Raman figure of merit (FOM) is a function of the Raman effective length, L_{eff}, the fiber effective area at the pump wavelength, $A_{eff\text{-}pump}$, and the Raman gain coefficient, g_R:

$$FOM = L_{eff} A_{eff\text{-}pump} / g_R$$

Depending on specific fiber type, a typical FOM is ~10 and has an effective length of ~20 km. The largest variability is the effective area at the wavelength of the pump, $A_{eff\text{-}pump}$, which varies between 40 and 80 μm^2.

4.2.8.6 OFA and Raman Amplifiers.
Table 4.1 provides a comparison between Raman and OFA amplification.

Table 4.1

Characteristics	Raman	OFA
Amplification band gain	Depends on pump offset, 25–50 nm per pump	Depends on dopant (Er, Th) ~20 nm
Gain	>10 dB	>20 dB
Gain tilt	Amplifies longer λs more than shorter, but it is adjustable; it may compensate for EDFA gain tilt	Amplifies longer λs more than shorter, but it is fixed
Output power	<30 dBm 500 mW	~20 dBm 100 mW
Noise	ASE, double Raleigh scatter	ASE
Noise figure	<1 dB	>5 dB
Flatness	1 dB	1 dB
Pump λ	<100 nm more than amplified range	980/1480 nm for Erbium
Pump power	~300 mW	~3 W
Saturation power	Depends on power of pump	Depends on dopant and gain; largely homogeneous saturation characteristics
Direction sense	Supports bidirectional signals	Unidirectional
Other	Possible cross talk among OChs due to OCh-to-OCh interaction	
Simplicity	Simpler	More complex

4.2.9 Stimulated Brillouin Scattering

Stimulated Brillouin scattering (SBS), in contrast to Raman scattering, is the nonlinear phenomenon by which a signal causes stimulated emission in both directions when a threshold power level is reached. However, emission in the same direction as the signal is scattered by acoustical phonons. Emission in the opposite direction is guided by the fiber but is downshifted by 11 GHz at 1550 nm (that is, it is of shorter wavelength than the signal); this is known as the *Brillouin frequency*. The downshifted SBS signal is also modulated as the signal is. Now, if another optical signal at the downshifted wavelength propagates in the same direction with the SBS generated signal, it will be added to it and cross talk takes place.

Experimentally, it has been determined that

- SBS is dominant when the spectral power of the source is high
- It abruptly increases when the launched power reaches a threshold value

Several factors determine the threshold value of launched power. Among them are the fiber material, the pump line width, the fiber length, the fiber-core effective cross-section area, and the bit rate of the signal. SBS threshold values are in the range of 5–10 mW of launched power (for externally modulated narrow line widths) and 20–30 mW for directly modulated lasers (see ITU-T G.663). Thus, Brillouin scattering, like Raman scattering, restricts the launched power per channel.

4.2.10 Wavelength Converters

Wavelength conversion is a critical function in DWDM systems and networks; it enables optical channels to be (spectrally) reallocated, adding to network flexibility and bandwidth efficiency. Wavelength conversion may be accomplished by properly managing the nonlinear properties of heterojunction semiconductor optical amplifiers or fibers, which otherwise have been deemed undesirable. Such properties

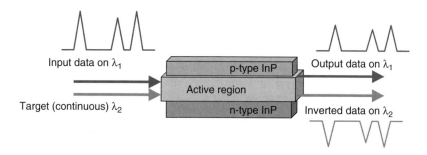

Figure 4.6. Cross-gain modulation transfers inverted data from one wavelength channel to another.

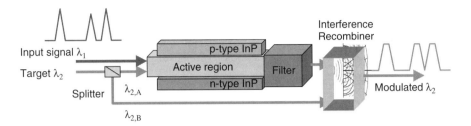

Figure 4.7. Principles of cross-phase modulating devices.

are *cross-gain modulation, cross-phase modulation, four-wave mixing,* and *optical-frequency shifting*.

4.2.10.1 Cross-Gain Modulation. When high optical power is injected into the active region and the carrier concentration is depleted through stimulated emission, *gain saturation* occurs and the optical gain is reduced. Based on this, two wavelengths are injected into the active region of a semiconductor optical amplifier; the modulated signal is at wavelength λ_1 and at high power, and wavelength λ_2, the target signal, is at lower power and is continuous (Figure 4.6). When the input bit in λ_1 is logic "one," the power is high and depletion occurs that blocks λ_2, hence, λ_2 is at logic "zero." When the bit in λ_1 is logic "zero" (no power), depletion does not occur and λ_2 passes at full power, hence, λ_2 is logic "one." Thus, a transfer of inverted data from λ_1 to λ_2 takes place. This method is known as cross-gain modulation (XGM).

4.2.10.2 Cross-Phase Modulation. The nonlinear properties of semiconductor materials can be used to take advantage of the phase difference induced on two propagating wavelengths. One wavelength (the signal), λ_1, is modulated and the other (the target), λ_2, is continuous. Consider also that λ_2 is split into two portions, $\lambda_{2,A}$ and $\lambda_{2,B}$ (Figure 4.7). Portion A is coupled in the nonlinear material and portion B bypasses it. However, as the modulated wavelength λ_1 enters the nonlinear material, it modulates its refractive index. As a result, the continuous "target" wavelength $\lambda_{2,A}$ is phase modulated. At the output of the nonlinear material, a filter rejects λ_1 and passes λ_{2-A}, which now recombines with λ_{2-B} and, through constructive/destructive interference, the two produce a recombined wavelength λ_2, which is modulated as the original signal λ_1, and, hence wavelength conversion is accomplished. This method is known as cross-phase modulation (XPM).

4.3 INTERACTION OF MULTIPLE CHANNELS IN FIBER

Interaction of multiple optical channels is a phenomenon in fiber-optic transmission known since the deployment of DWDM technology. Prior to DWDM, transmission was at 1310 and/or 1550 nm and, thus, channel interaction was negligible or nonexistant. Optical channels interact because of the nonlinear behavior of the transmission medium via photon–matter–photon interaction. We already have described such interactions in previous sections on fiber amplification and wavelength converters. In this section, we discuss some more nonlinear phenomena as they affect the quality of signal.

4.3.1 Nonlinear Phenomena

Optical channel interactions caused by nonlinear phenomena are best described by quantum theory and, thus, we only provide a quantitative description here. They are distinguished as *forward scattering* and *backward scattering* (Raman and Brillouin scattering, discussed previously) and as *four-wave (or four-photon) mixing*. The direction (forward and backward) is with respect to the direction of the excitation. In DWDM systems, nonlinear phenomena are viewed as both advantageous and as degrading:

- *Advantageous*, because lasers, optical amplifiers and dispersion compensation are based on them.
- *Degrading*, because signal losses, noise, cross-talk, and pulse broadening are caused by them.

The behavior of a dielectric with respect to optical power is like that of a dipole. It is the dipole nature of the dielectric that interacts harmonically with electromagnetic waves. Thus, the polarization of an electromagnetic wave, P, induced in the electric dipoles of a medium by an electric field, E, is proportional to *susceptibility*, χ:

$$P = e_0[\chi^1 \cdot E + \chi^2 \cdot E \cdot E + \chi^3 \cdot E \cdot E \cdot E + \ldots]$$

where e_0 is the permitivity of free space.

When the optical power is low, it results in small oscillations and the first term of the series approximates the photon–fiber system behavior (i.e., a linear system). However, when the optical power is large, then the oscillations are such that higher-order terms (nonlinear behavior) become significant.

For an *isotropic* medium, the first term represents the linear behavior of matter, whereas the second-order term is orthogonal and thus vanishes (or is negligible). Silica glass, unlike quartz crystals, is amorphous with macroscopic inversion symmetry. Thus, it has zero (or near-zero) second-order nonlinearity. However, the third-order term produces nonlinear effects that can be significant, such as stimulated Raman scattering (SRS), stimulated Brillouin scattering (SBS) (already discussed), and four-wave mixing (FWM).

4.3.2 Four-Wave Mixing

Consider three lightwave frequencies, f_1, f_2, and f_3, closely spaced. Then, from the interaction of the three, a fourth lightwave frequency is generated, f_{fwm}, such that

$f_{fwm} = f_1 + f_2 - f_3$. This is known as *four-wave mixing* (FWM), or *four-photon mixing*. The order of lightwave frequencies is $f_1, f_{fwm}, f_3,$ and f_2. In fact, ITU-T defines FWM components as $f_{fwm} = f_1 \pm f_2 \pm f_3$.

The FWM component, E_{fwm}, generated by three components E_1, E_2, and E_3, at the output of a fiber segment, L, with angular frequency ω, refractive index n, nonlinear refractive index χ, and loss α is described by

$$E_{fwm} = j[(2\pi\omega)/nc]d\chi E_1 E_2 E_3 e^{-\alpha(L/2)} F(\alpha, L, \Delta\beta)$$

where $F(\alpha, L, \Delta\beta)$ is a function of fiber loss, fiber length, and propagation variation (phase mismatch), related to channel spacing and dispersion.

The output power of the f_{fwm} and the efficiency of the four-wave mixing mechanism depends on

- The wavelength mismatch or channel spacing, Δf
- The power intensity of the contributing frequencies $f_1, f_2,$ and f_3
- The fiber chromatic dispersion
- The refractive index
- The fiber length
- Higher-order polarization properties of the material (nonlinear Kerr coefficient)

The effect of four-wave mixing on optical transmission and in single-mode fiber is

- Cross talk, due to superposition of uncorrelated data from the contributing channels
- Signal-power depletion as a result of signal-power sharing among the contributing channels to the FWM-generated channel
- Signal-to-noise degradation, due to superposition of noise and random data from the contributing frequencies

As the signal-input power of $f_1, f_2,$ and f_3 increases, or as the channel spacing decreases, the FWM term f_{fwm} increases. ITU-T (G.663) recommends that the critical optical power for FWM be greater than 10 mW, although this depends on channel spacing. At 200 Ghz channel spacing, it has been experimentally verified that the FWM effect is drastically decreased as compared with 100 GHz or less.

As the channel spacing decreases to 25 GHz, FWM cross talk becomes more significant. A widely used formula for FWM induced cross talk is given by

$$P_{ijk}(L) = (\eta/9)D^2\gamma^2 P_i P_j P_k e^{-\alpha L}\{[1 - \exp^{-\alpha L}]^2/\alpha^2\}$$

where P_i, P_j, and P_k are the input powers of the three input signals $f_1, f_2,$ and f_3, L is the optical traveled length (i.e., fiber length), α is the attenuation coefficient, η is the FWM efficiency, D is a degenerative factor (equal to 3 for degenerative FWM or 6 for nondegenerative FWM), and γ is a nonlinear coefficient (for the fiber medium) given by

$$\gamma = (2\pi n_2)/(\lambda A_{\text{eff}})$$

where λ is the wavelength in free space, and A_{eff} and n_2 is the effective area and the nonlinear refractive index of the fiber, respectively.

The nonlinear refractive index is related to nonlinear susceptibility χ_{1111} of the optical medium and to the refractive index of the core n of the fiber, which is expressed as

$$n_2 = [(48\pi^2)/(cn^2)]\,\chi_{1111}$$

The FWM efficiency factor depends on, among other parameters, a phase-matching factor, which also depends on fiber dispersion and channel spacing. In fact, at the zero-dispersion wavelength, FWM exhibits its best performance; this is one of the reasons why in DWDM communications, the zero-dispersion wavelength is avoided and the wavelength operating point is preferably slightly above or below the zero-dispersion wavelength. Thus, four-wave mixing, as opposed to Raman scattering, requires strong phase matching of coincident energy from all three wavelengths. Moreover, both chromatic dispersion and length of fiber reduce the intensity of the FWM product.

In short, FWM efficiency depends on material dispersion, channel separation, optical traveled path (or fiber length), and optical power level of each contributing channel. The resultant FWM component affects the quality of the signal in terms of cross talk and bit error rate and, thus, FWM limits the channel capacity of a fiber system.

4.3.3 Temporal FWM

Consider a narrow light pulse at a wavelength traveling along a fiber segment. Think of the light pulse as a window sliding along the fiber at a constant speed. As the light pulse slides along the fiber, it influences the electric dipoles of the fiber segment for as long as it is in that segment. If at the same time, there are two more pulses on adjacent wavelength channels, overlapping in time, then in this segment four-wave mixing occurs. Since the signal power at the *near end* of the fiber (close to the source) is at its maximum, the four-wave mixing product is also at its highest intensity. This is a consequence of the FWM efficiency factor and the strong phase matching required for FWM.

Similarly, when the three light pulses arrive at a *far-end* segment of a several-kilometers-long fiber, due to dispersion and attenuation of the pulses, the four-wave mixing product in that segment is at its lowest intensity. Figure 4.8 illustrates qualitatively the contribution of FWM at the near end and at the far end of a fiber. Clearly, FWM occurs in a time-continuous manner, and although at the near end it has a

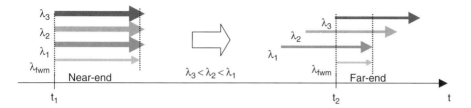

Figure 4.8. The effect of FWM is strongest at the near end of synchronized channels, and it is diminished at the far end, where channels are weakest due to attenuation and dispersion.

maximum effect, its cumulative strength is diminishing as the pulses travel along the fiber. The study of FWM in the time domain is termed *temporal FWM* (tFWM).

4.3.4 Impact of FWM on DWDM Transmission Systems

As the density of wavelengths (channels) in DWDM systems increases, then cross talk noise increases in importance. Thus,

- As the optical power of each channel is increased, four-wave mixing becomes more intense.
- As the launched optical power of each channel is lowered, then the actual fiber length is decreased to assure that the arriving signal can be detected reliably. This may necessitate optical amplification to extend the fiber path (optical amplification increases the cost).
- As the channel (wavelength) density increases, or if the channels are spaced very close to each other, four-wave mixing becomes more intense.
- As the channels are spaced further apart, then fewer wavelength channels can fit in the fiber.

4.3.5 Summary of Nonlinear Phenomena

In summary, the most important nonlinear phenomena in optical communication are

- Stimulated Raman scattering (SRS). OChs with optical power above a threshold may behave as a pump for longer wavelengths or cause photonic spontaneous emission in either direction, which degrades the signal-to-noise ratio. There is no well-known controlling mechanism for SRS.
- Stimulated Brillouin scattering (SBS). This is similar to SRS: dominant in the backward direction and with threshold is from ~5 to 10 mW for external modulation and from ~20 to 30 mW for direct modulation. SBS is controlled by lowering the signal intensity, or by making the source linewidth wider than the Brillouin bandwidth.

- Four-wave mixing (FWM). Created sidebands deplete the optical power of contributing signals. FWM is controlled with channel spacing and optical power management.
- Modulation instability (MI). MI degrades the signal-to-noise ratio due to created sidebands and thus decreases the optical power level.
- Self-phase modulation (SPM). Changes in optical intensity cause variation in the phase of the optical signal; this broadens the signal spectrum (and pulse width). Operating in the anomalous dispersion region, chromatic dispersion and SPM compensate each other. However, this may result in spontaneous formation of solitons.
- Cross-phase modulation (XPM). Interacting adjacent optical channels induce phase changes and thus pulse broadening. This is controlled by channel spacing selection.

4.3.6 Factors Affecting Matter and Light

We have discussed the parameters that influence light and matter and also the interactions that take place between light and matter. In Table 4.2, we summarize what causes these effects.

4.4 REVIEW OF DWDM NETWORKS

New multiservice applications, distinct from traditional ones, have created a bandwidth demand that only the DWDM optical network can meet. For example, streaming MP3 requires 200 Kbit/s, high-quality videoconferencing requires 3

Table 4.2. Cause and Effect

Cause	Effect
λ interacts with λ	Interference (coherent superposition of states)
λs interact with matter	Linear and nonlinear effects: absorption, scattering, birefringence, phase shift, reflection, refraction, diffraction, polarization, polarization shift, PDL, modulation, self-phase modulation, etc.
λ–matter–λ interaction	FWM issues, SRS, SBS, OFA
Nonmonochromatic channel	Pulse broadening, finite number of channels within available band
Refractive index variation (n)	Affects propagation of light
Transparency variation	Affects amount of light through matter;
Scattering	optical power loss (attenuation)
Reflectivity	Affects polarization of reflected optical wave; affects phase of reflected optical wave
Ions in matter	Dipoles interacting selectively with λs; Energy absorption or exchange; affect refractive index

Mbit/s, and two-channel DVD-quality streaming video requires 10–14 Mbit/s. The evolving DWDM technology requires careful deployment in an optical network that offers multiple services reliably and meets service-quality metrics and performance parameters. Optical-network nodes are interconnected into a well-managed transparent network, which is remotely provisioned, quickly reconfigured, easily maintained, cost-effective, and quickly restored (less than 50 ms). This optical network is intelligent, robust, flexible, and scalable, more resilient, and fault, tolerant and it delivers multiservices with end-to-end service provisioning.

The next-generation DWDM network will support both established traditional services and new differentiating services such as real-time multicast, streaming video, e-commerce, and virtual and virtual private networks (VPNs). This means that overlay networks will converge to a single and more potent optical network. However, such a network demands that certain parameters of the data network such as quality of service (QoS) and "best effort" transport be revised. And this revision will require:

- Protocols to support QoS, such as IP differentiated services (DiffServ)
- Protocols to support traffic engineering with dynamic connectivity in the packet/cell-based network, such as multiprotocol label switching (MPLS).
- Protocols to support bandwidth as well as wavelength management with dynamic connections at the optical layer, such as multiprotocol lambda (wavelength) switching (MPλS).
- Fast, transparent, and multiservice optical switches
- Flexible, adaptable, and intelligent edge or access nodes
- Interoperation of both optical core and edge network elements
- Dynamic optical signaling
- Advanced network-management systems
- Security

As the number of wavelengths per fiber increases, and as the bit rate per wavelength increases and as new services requiring more bandwidth are offered, core services come closer to the edge. This also brings fiber closer to the edge and closer to the home/office. As a consequence, the number of network layers decreases, and the edge or access nodes become smarter and able to support a variety of protocols and bit rates, scalability with a fine granularity of bandwidth traffic, provisionability, reliability, availability, fault tolerance, restoration, and intelligence. This will enable us to provide reliable services not only on the wavelength level but also on elastic granularity that ranges from DS1/E1 to GbE over *optical virtual private networks* (OVPN). These networks will also enable bandwidth trading and reselling (bandwidth on demand), multiservices, and service transparency with various *service-level agreements* (SLAs). SLA metrics pertain to

- Information rate
- Network availability (this is not exactly the same as network down-time)

- Latency
- Performance-level parameters
- Quality of Service
- Time to report
- Time to respond
- Compensation
- On-time provisioning

Although new fiber networks are currently in deployment or in planning, DWDM networks require additional efficiency requirements that are related to demographics, communications customs, people's habits and trends, economic profiles, and embedded network technologies that will influence the future network.

Among the new protocols designed for deployment in the next-generation DWDM optical network is the generalized MPLS (GMPLS), the generic framing procedure (GFP), the *Link Capacity Adjustment Scheme* (LCAS), the optical transport network (OTN), and several new routing and wavelength-assignment algorithms. Research continues to improve existing and newly developed algorithms; management protocols and standards have been issued recently and many proposals are under study and evaluation.

4.4.1 Next-Generation SONET/SDH

The Generic Framing Procedure (GFP) is a relatively new ITU-T standard (G.7041/Y.1303, 2001) that defines a flexible encapsulation framework for traffic adaptation of broadband transport applications and it supports client-control functions that allow different client types to share a channel. In addition, the GFP provides an efficient mechanism to map broadband data protocols (such as Fibre Channel, ESCON, FICON, GbE) onto multiple concatenated STS-1 payloads in a SONET/SDH frame. This mapping of a physical layer or logical layer signal to a byte-synchronous channel supports different network topologies with low latency of packet-oriented or block-coded data streams, with differentiated QoS features meeting service-level agreement (SLA) requirements. The flexible encapsulation of diverse protocols onto GFP-generalized frames, and mapping of GFPs onto the SPE of SONET/SDH frames to support both packet and circuit switching services is what differentiates the Next Generation SONET/SDH (NG-S) from legacy SONET/SDH. Thus, GFP allows for existing circuit switching, SONET, GbE, and other packet-based protocols to be used as an integrated and interoperable transport platform that provides cost efficiency, QoS, and SLA as required by the client. In the following, we take a closer look at the GFP.

The motivation for the NG-S is to provide a standardized, robust, and efficient method of transporting all types of data in addition to synchronous traffic (e.g., voice and real-time video). The Generic Framing Procedure (GFP) is believed to accomplish this by encapsulating GbE, IP, FC, and others and transporting virtual concatenated SONET/SDH frames and LCAS over the next-generation SONET/SDH (NG-S) (Figure 4.9).

The *Link Capacity Adjustment Scheme* (LCAS) allows containers to be added or removed dynamically in the NG-S payload. Dynamic addition/deletion of bandwidth is also used for load balancing. LCAS is accomplished using control packets to configure the path between source and destination.

The next-generation SONET/SDH is defined to support two platforms: a multiservice provisioning platform (MSPP) and a multiservice switching platform (MSSP). The MSPP provides aggregation, grooming, and switching capabilities. It responds to alarm and error SONET/SDH conditions, and it supports new protection schemes for many topologies. The MSSP provides bandwidth and wavelength management via large, nonblocking switching fabrics (cross-connects).

4.4.2 Topologies

The most well-known optical network topologies are the ring, the all-connected mesh, and point-to-point. The two-fiber Bi-directional Line Switched Ring (2f-BLSR) is applicable to small- and medium-size LANs and Metropolitan ring networks (Metro). It consists of a dual-fiber counterrotating ring. In this scheme, half

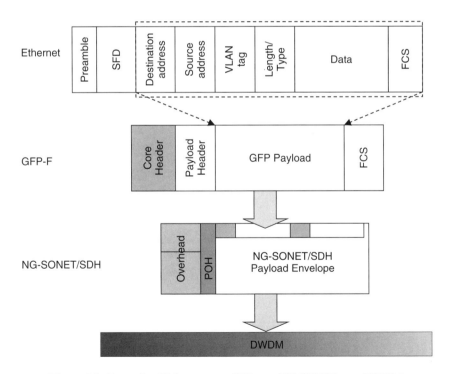

Figure 4.9. Example of Ethernet over GFP over NG-SONET over DWDM.

of the bandwidth on the ring is allocated to protection. Thus, if each ring uses an OC-48, STS-1 to STS-24 are used for traffic and STS-25 to STS-48 are allocated for protection. When a single fiber cut occurs, the traffic is routed over the healthy ring and it uses the allocated available bandwidth, in which case all STSs are used, STS-1 to 48. Switch to protection complies with SONET standards (<50 ms).

The four-fiber Bi-directional Line Switched Ring (4f-BLSR) is applicable to medium- and large-size LANs and Metropolitan ring networks (Metro). It consists of two dual-fiber rings with counterrotating rings. In this scheme, the 4f-BLSR combines both ring and span protection. In WDM 4f-BLSR rings, coarse (20 with 400 GHz separation) or dense wavelengths (40 with 100 GHz separation) are used. The typical rate per wavelength is in the range OC-48 to OC-192. Switch to protection complies with SONET standards (<50 ms).

Next-generation optical rings combine all the features of existing rings and increase functionality and intelligence using MSPP nodes, whereby multiple rings are interconnected with MSSP hub nodes (Figure 4.10).

The Mesh network consists of cross-connecting nodes. Some nodes are O–E–O and some all optical. Large nodes are capable of carrying aggregate traffic at extremely high capacities, and also can add–drop and groom traffic. Each interconnecting link consists of many dual fibers (fiber per direction), and they include signal conditioners (equalization and regeneration) when the optical signal is transported over distances hundreds of kilometers long. The mesh network has excellent protection strategies, as nodes may be reprovisioned to reroute traffic away from the failure condition (fiber cut or failed node). Faults are detected with power detectors and performance-parameter thresholds. Reconfiguration may be autonomous by switching to protection to SONET/SDH standards (switch to protection takes less than 50 ms). Reconfiguration may also be via network management procedures. More sophisticated management protocols perform reconfiguration to

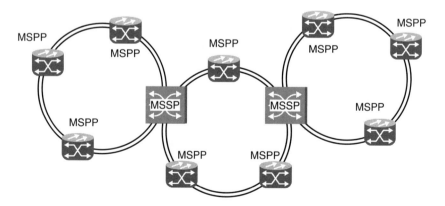

Figure 4.10. Next-generation optical ring networks interconnected with MSSP nodes.

achieve traffic balancing and traffic grooming. The Path Protected Mesh Network (PPMN) may be viewed as consisting of many next-generation optical rings interconnected with large MSSPs that are capable of carryingaggregate traffic at extremely high capacities (Figure 4.11).

The network elements of the PPMN are MSPPs capable of aggregating a variety of client data, including voice/video, and can add–drop and groom traffic. Each interconnecting link consists of few or many dual fibers (fiber per direction), and signal transport may depend on signal conditioners (equalization and regeneration) to be able to transport the optical signal over distances of hundreds of kilometers long.

The *point-to-point* topology establishes connectivity between two users. In a strict sense, it consists of two transceivers and an optical link or path between them, hence its name. Point-to-point is applicable in a range of fiber lengths, from a few hundred meters to many thousands of kilometers. However, these two extremes address different applications and require entirely different engineering effort and cost structure. Therefore, point-to-point networks are distinguished as short-haul (a few kilometers long), medium-haul (a few tens of kilometers long), long-haul (a few hundreds of kilometers long) and ultra long-haul (a few thousands of kilometers). In addition, there is the very short-haul (a few hundreds of meters) and ultrashort-haul for distances of few meters or even decimeters.

4.4.3 The Optical Transport Network

The *optical transport network* (OTN) is relatively a new ITU-T recommendation (G.709, G.872, and G.959) that was developed for long-haul transport at data rates from 2.5 to 40 Gbit/s.

OTN, like the NG-S, defines a synchronous payload within a fixed frame length, a comprehensive overhead to support a variety of client payloads, and forward error correction (FEC) to provide more margin for fixed fiber span or increased fiber span at BER objectives that can be met.

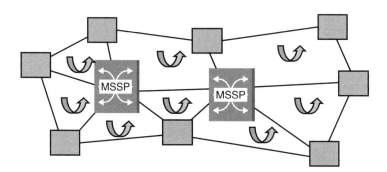

Figure 4.11. Next-generation optical mesh networks: multiple optical rings interconnected with MSSP nodes.

4.4.4 DWDM Network Restoration

An important aspect of opticaql networks is how fast a network restores service. The total interval from fault detection to service restoration should be short so that service is perceived to be uninterrupted (in SONET/SDH this interval has been set to 50 ms). However, in high-bandwidth DWDM networks, even a short interval may seem too long if aggregate traffic is higher than a Tbit/s per fiber. For example, at one Tbit/s aggregate traffic (such as in backbone or ultra long-haul networks), a 50 ms interruption corresponds to 50 billion bits lost. However, at 1 Gbit/s (such as in small metro and access networks), a 50 ms interruption corresponds to 50 million bits lost. Therefore, restoration time performance need not be the same for all network layers (Figure 4.12).

However, DWDM networks not only deal with fiber and node degradation/faults but also with optical channel degradation/faults, either due to hard faults (laser, detector) or linear and nonlinear interchannel interaction (see Chapters 1 and 2). In DWDM, even degradations may be significant, particularly at ultrahigh bit rates. DWDM channels at 40 Gbit/s, when degraded beyond an acceptable limit, should require restoration. Therefore, service protection in DWDM encompasses the following key functions and features:

- Comprehensive degradation/fault monitoring
- Efficient and accurate degradation/fault detection, correlation, and localization
- A reporting mechanism to a fault manager

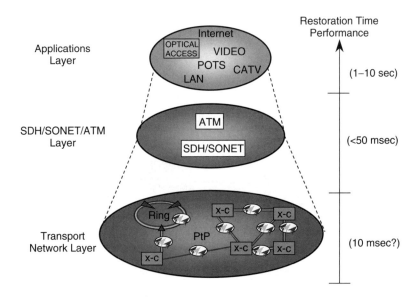

Figure 4.12. Restoration time in layered networks.

- Efficient and reliable signaling to notify other downstream and upstream nodes
- Efficient routing algorithm and bandwidth management
- Network architecture with fast restoration schemes
- Efficiently provisionable nodes
- Network with built-in overcapacity
- Efficiently and well balanced inter- and intranetworking interfaces

Similarly, the performance of each network topology depends on many factors. Among them are:

- Number of nodes
- Network scalability, elasticity, and flexibility
- Seamless evolution
- Maximum traffic capacity
- Service restoration and disaster-avoidance capability
- Quality of service and cost of service
- Real-time aspects of supported traffic
- Variability of supported traffic
- Reliability
- Network resilience
- Network response and agility
- Number of fiber links between nodes
- Trends and market preferences (for example, a ring system may be preferred because of familiarization with SONET/SDH embedded networks, or a mesh system may be preferred because of better survivability capability)

Any of the topologies already described provide protection and restoration capabilities. These are on the link level, on the channel level, on the client payload level, on the multiplex level, and so on. ITU-T (G.872) specifies identifiers in the frame overhead for optical-channel operation, administration and management (OCh-OAM) that are used to localize faults, isolate faults, and provide a mechanism for automatic protection switching:

- *Trail Trace* (OCh-ID). Used to verify connectivity through OADMs, optical cross-connects, fiber-patch panels, and so on.
- *Signal Label* (OCh-SL). Used to verify that the client signal is compatible with the equipment.
- *Per-channel automatic protection switching* (OCh-APS). Used for OCh Shared Protection Rings (OCh/SPRINGs) and OCh Trail Protection.
- *Forward and backward defect indication* (OCh-FDI/BDI). Used to localize faults and to enable single-ended maintenance.

- *Per-channel signal quality monitoring and backward quality indication* (OCh-SQM/BQI). Used to verify quality of service and as an instrument to isolate sources of degraded performance.
- *Per-channel tandem connection maintenance* (OCh-TCM). Used to maintain a channel through an entire subnetwork that does not include the OCh termination elements

4.4.5 Bandwidth Management

Bandwidth management is a function required by all systems and networks, voice or data, small or large, so that client's data is delivered with the agreed-upon quality. As traffic becomes more elastic, the aggregate bandwidth keeps growing, and as faults and severe degradations occur on nodes, fiber links, and optical channels, traffic management becomes more important and additional requirements are added to increase efficiency.

4.4.6 Wavelength Management

In DWDM systems and networks where each wavelength is used as a separate channel, it is reasonable to consider the *failure in time* (FIT) of the various optical components on the path. When a component fails, it is considered a hard failure and it should be easily detected. However, there may be severe degradations that impact the quality of signal so that an optical channel performs below an acceptable level (for example, BER $< 10^{-9}$), and yet may not be detected. Although monitoring for degradations is more complex, certain severe degradation mechanisms are as important as hard failures. In such cases, when the quality of the optical signal becomes unacceptable, the channel should be dynamically switched to protection in the same fiber or in another.

In traditional single-wavelength networks, wavelength collision clearly has not been an issue since the optical channel operates at only one wavelength (1310 nm). Thus, moving this channel from one fiber to another it was a straightforward operation, as long as a protection fiber was available. In DWDM networks, this is not as straightforward for many reasons. In fact, this may lead to a new possible situation that we term *wavelength collision*. Here, we define *wavelength collision* as the action by which one optical channel associated with a particular wavelength is switched from fiber A to fiber B, where in fiber B the same wavelength is already in use; this definition is extendable to more than one wavelength. Figure 4.13 illustrates a mesh network requiring wavelength conversion to avoid wavelength collision. However, these systems and networks have several attractive attributes:

- Optical channels allow for format and bit rate independence flexibility
- Wavelength conversion minimizes stranded (available) capacity
- Optical channels are wavelength independent

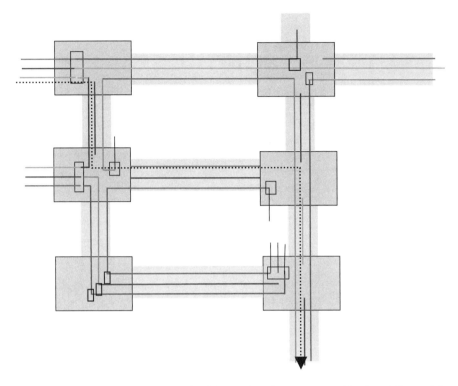

Figure 4.13. A DWDM mesh network with wavelength management (the dotted-line path is illustrative of the wavelength conversions).

- Systems support easy access to the optical layer
- Systems support multivendor environment
- Systems support scalability and upgradability
- Systems support fast fault detection and fault localization
- Systems support real-time communication and QoS

4.4.7 Optical-Performance Monitoring

Performance monitoring is a critical function performed at the input of each signal at each node, if the signal is in the electronic regime. However, today, there is no solution for performance monitoring in the optical regime, although research on this subject continues and possible solutions have been proposed. Therefore, performance monitoring and error correction is currently limited to opaque systems that are strategically positioned at the edge of network domains (Figure 4.14).

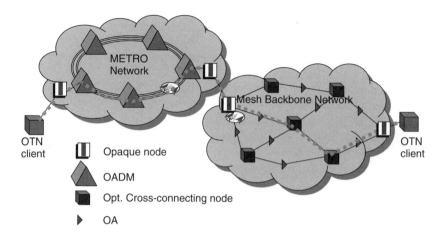

Figure 4.14. Opaque nodes at the edges of all-optical DWDM networks provide PM and EDC functionality.

REFERENCES

1. T. Wildi, *Units and Conversion Charts,* 2nd ed., IEEE Press, 1995.
2. G.J. Zissis, *"The Infrared Handbook,"* ERIM and SPIE, 1978.
3. W.L. Wolfe, *"Introduction to Infrared System Design,"* SPIE Press, 1997.
4. K. Nassau, *The Physics and Chemistry of Color,* Wiley, 1983.
5. S.V. Kartalopoulos, *DWDM: Networks, Devices and Technology,* Wiley/IEEE Press, 2003.
6. S.V. Kartalopoulos, *Fault Detectability in DWDM: Towards Higher Signal Quality and Network Reliability,* IEEE Press, 2001.
7. S.V. Kartalopoulos, *Introduction to DWDM Technology: Data in a Rainbow,* IEEE Press, 2000.
8. B. Mukherjee, *Optical Communication Networks,* McGraw-Hill, 1997.
9. I.P. Kaminow and T.L. Koch (Eds.), *Optical Fiber Communications IIIA* and *Optical Fiber Communications IIIB,* Academic Press, 1997.
10. S.O. Kasap, *"Optoelectronics and Photonics,"* Prentice-Hall, 2001.
11. H. Kogelnik, "Theory of Dielectric Waveguides," in *Integrated Optics,* 2nd ed., Chapter 2, T. Tamir (Ed.), Springer-Verlag, 1979.
12. G.P Agrawal, *Nonlinear Fiber Optics,* 2nd ed., Academic Press, 1995.
13. P. Yeh, "Introduction to Photorefractive Nonlinear Optics," Wiley, 1993.
14. S.V. Kartalopoulos and P. Verma, "Pragmatic Teaching of Advanced Optical Networks: Connecting Physics, Optical technology and Networks," in *Proceedings of the IEEE International Conference on Information Technology: Coding and Computing* (ICTT 2003), pp. 65–68, Las Vegas, 28–30 April 2003.

15. S.V. Kartalopoulos, "Ultra-High Bandwidth Data Transport for DWDM Short, Medium-Haul and Metro Using Low Bit Rates," in *APOC 2002 Conference, Optical Switching and Optical Interconnection II,* pp. 100–107, October 2002, Shanghai, China.
16. S.V. Kartalopoulos, "Consumer Communications in the Next Generation Access Network," in *CCNC-2004 Conference,* Las Vegas, January 5–8, 2004, session N13, paper no. 0-7803-8146-7.
17. S.V. Kartalopoulos, "Ultra-fast Pattern Recognition in Broadband Communications Systems," in *ISPACS'98,* November 1998, Melbourne, Australia, pp. 558–591.
18. R.H. Stolen, "Non-Linear Properties of Optical Fibers," in *Optical Fiber Telecommunications,* S.E. Miller and G. Chynoweth (Eds.), Academic Press, 1979.
19. L. Desmarais, *Applied Electro-Optics,* Prentice-Hall, 1999.
20. S.V. Kartalopoulos, "Bandwidth Elasticity with DWDM Parallel Wavelength-bus in Optical Networks," *SPIE Optical Engineering, 43,* 5, May 2004, pp. 1092–1100.
21. A. Yariv, "Coupled Mode Theory for Guided Wave Optics," *IEEE Journal Quantum Electron., QE-9,* 1973, pp. 919–933.
22. J.B. Eom, D.S. Moon, U-C. Paek, and B.H. Lee, "Fabrication and Transmission Characteristics of Couplers Using Photonic Crystal Fibers," *Technical Digest OFC 2002,* paper ThK-2, pp. 465–466, March 2002.
23. U. Cummings and W. Bridges, "Bandwidth of Linearized Electro-Optic Modulators," *Journal of Lightwave Technology, 16,* 1482–1490, August 1998.
24. C.R. Giles and M. Spector, "The Wavelength Add/Drop Multiplexer for Lightwave Communication Networks," *Bell Labs Technical Journal, 4,* 1, 207–229, 1999.
25. W.K. Chen, *Passive and Active Filters,* Wiley, 1986.
26. G. Hernandez, *Fabry-Perot Interferometers,* Cambridge University Press, 1986.
27. J. Stone and D. Marcuse, "Ultrahigh finesse fiber Fabry-Perot interferometers," *Journal Lightwave Technology, LT-4,* 4, 382–385, April 1986.
28. A.A.M. Saleh, and J. Stone, "Two-Stage Fabry-Perot Filters as Demultiplexers in Optical FDMA LAN's," *Journal Lightwave Technol., LT-7,* 323–330, 1989.
29. A. Iocco, H.G. Limberger, and R.P. Salathe, "Bragg Grating Fast Tunable Filter," *Electr. Lett., 33,* 25, 2147–2148, December 1997.
30. S.R. Mallison, "Wavelength-Selective Filters for Single-Mode Fiber WDM Systems Using Fabry-Perot Interferometers," *Applied Optics, 26,* 430–436, 1987.
31. H. Kobrinski and K.W. Cheung, "Wavelength-Tunable Optical Filters: Applications and Technologies," *IEEE Commun. Mag., 27,* 53–63, October, 1989.
32. R. Tewari and K. Thyagarajan, "Analysis of Tunable Single-Mode Fiber Directional Couplers Using Simple and Accurate Relations," *Journal Lightwave Technology, LT-4,* 4, 386–399, 1986.
33. R.M.A. Azzam and N.M. Bashara, "*Ellipsometry and Polarized Light,*" North Holland, 1977.
34. P. Lu, L. Chen, and X. Bao, "Polarization Mode Dispersion and Polarization Dependent Loss for a Pulse in Single-Mode Fibers," *Journal Lightwave Technology, 19,* 6, 856–860, June 2001.
35. P. Lu, L. Chen, and X. Bao, "Principal States of Polarization for an Optical Pulse in the Presence of Polarization Mode Dispersion and Polarization Dependent Loss," in *Pro-*

ceedings 2000 International Conference Applications Photonic Technology (ICAPT'2000), Quebec, Canada, June 12–16, 2000.
36. N. Gisin, "Statistics of polarization dependent losses," *Optics Communications, 114,* 1995, pp. 399–405.
37. K. Shiraishi, S. Sugayama, and S. Kawakami, "Fiber Faraday Rotator," *Applied Optics, 23,* 1103, 1984.
38. W. Eickhoff, "n-Line Fiber-Optic Polarizaer," *Electronics Letters, 6,* 762, 1980.
39. T. Hosaka, K. Okamoto, and J. Noda, "Single-Mode Fiber Type Polarizer," *IEEE Journal Quantum Electronics, QE-18,* 1569, 1982.
40. R.A. Bergh, H.C. Lefevre, and H.J. Shaw, "Single-Mode Fiber-Optic Polarizer," *Optics Letters, 5,* 479, 1980.
41. J.P. Gordon, and H. Kogelnik, "PMD fundamentals: Polarization mode dispersion in optical fibers," *Proceedings of the National Academy of Sciences, 97,* 9, 4541–4550, April 25, 2000.
42. C.D. Poole, R.W. Tkach, A.R. Chraplyvy, and D.A. Fishman, "Fading in Lightwave Systems Due to Polarization-Mode Dispersion," *IEEE Photonics Technolology Letters, 3,* 68–70, January 1991.
43. E. Lichtman, "Limitations Imposed by Polarization-Dependent Gain and Loss on All-Optical Ultralong Communications Systems," *Journal Lightwave Technol., 13,* 906–913, May 1995.
44. N. Shibata, K. Nosu, K. Iwashita, and Y. Azuma, "Transmission Limitations Due to Fiber Nonlinearities in Optical FDM Systems," *IEEE Journal on Selected Areas in Communications, 8,* 6, 1068–1077, August 1990.
45. D. Cotter, "Stimulated Brillouin Scattering in Monomode Optical Fiber," *Journal Opt. Commun., 4,* 10–19, 1983.
46. G. Waarts and R.P. Braun, "Crosstalk Due to Stimulated Brillouin Scattering In Monomode Fibers," *Electronics Letters, 24,* 78–80, 1988.
47. A.M. Hill, and D.B. Payne, "Linear Crosstalk in Wavelength Division Multiplexed Optical Fiber Transmission Systems," *Journal Lightwave Technol., LT-3,* 643–651, 1985.
48. P.T. Thomas et al., "Normal Acoustic Modes and Brillouin Scattering in Single-Mode Optical Fibers," *Physics Revues, B19,* 4986–4998, 1979.
49. D. Nesset, T. Kelly, and D. Marcenac, "All-Optical Wavelength Conversion Using SOA Nonlinearities," *IEEE Communications Magazine, 36,* 12, 56–61, November 1998.
50. S. Sudo (Ed.), *Optical Fiber Amplifiers: Materials, Devices, and Applications,* Artech House, 1997.
51. G. Eisenstein, "Semiconductor Optical Amplifiers," *IEEE Circuits and Devices Magazine, 5,* 4, 25–30, July 1989.
52. A.R. Chraplyvy, "Optical Power Limits in Multi-Channel Wavelength-Division-Multiplexed System Due to Stimulated Raman Scattering," *Electronics Letters, 20,* 2, 58–59, January 1984.
53. K. Vilhelmsson, "Simultaneous Forward and Backward Raman Scattering in Low-Attenuation Single-Mode Fibers," *Journal Lightwave Technology, LT-4,* 4, 400–404, 1986.

54. N.A. Olsson and J. Hegarty, "Noise Properties of a Raman Amplifier," *IEEE Journal of Lightwave Technology, LT-4,* 4, 396–399, April 1986.
55. J.R. Thompson and R. Roy, "Multiple Four-Wave Mixing Process in an Optical Fiber," *Optics Letters, 16,* 8, 557–559, April 1991.
56. K. Inoue, "Four-Wave Mixing in an Optical Fiber in the Zero-Dispersion Wavelength Region," *IEEE Journal Lighwave Technology, LT-10,* 11, 1553–1563, November 1992.
57. K. Inoue, "Suppression of Fiber Four-Wave Mixing in Multichannel Transmission Using Birefringent Elements," *IEICE Transactions on Communications, E76-B,* 9, 1219–1221, Sept. 1993.
58. K. Inoue, "Fiber Four Wave Mixing in Multi-Amplifier Systems with Nonuniform Chromatic Dispersion," *IEEE Journal Lighwave Technology, LT-13,* 82–93, Jan. 1995.
59. J.P. Pocholle, J. Raffy, M. Papuchon, and E. Desurvire, "Raman and Four Photon Mixing Amplification in Single Mode Fibers," *Optical Engineering, 24,* 4, 600–608, 1985.
60. S. Song, C.T. Allen, K.R. Demarest, and R. Hui, "Intensity-Dependant Phase-Matching Effects on Four-Wave Mixing in Optical Fibers," *Journal of Lightwave Technology, 17,* 11, 2285–2290, November 1999.
61. J. Hansryd, H. Sunnerud, P.A. Andrekson, and M. Karlsson, "Impact of PMD on Four-Wave-Mixing-Induced Crosstalk in WDM Systems," *IEEE Photonics Technology Lett., 12,* 9, 1261–1263, Sept., 2000.
62. R.F. Slusher, L.W. Hollberg, B, Yurke, J.C. Mertz, and J.F. Valley, "Observation of Sqeezed States Generated by Four Wave Mixing in an Optical Cavity," *Phys. Rev. Lett., 55,* 2409–2412, 1985.
63. A. Evans, "Raman Amplification in Broadband WDM Systems," *Technical Digest OFC 2001,* paper TuF4-1, March 2001.
64. C.R.S. Fludger, V. Handerek, and R.J. Mears, "Fundamental Noise Limits in Broadband Raman Amplifiers," *Technical Digest OFC 2001,* paper MA5-1, March 2001.
65. M.Bolshtyansky, J. DeMarco, and P. Wysocki, "Flat, Adjustable Hybrid Optical Amplifier for 1610nm-1640nm band," *Technical Digest OFC 2002,* paper ThJ5, pp. 461–462, March 2002.
66. H.Masuda, "Review of Wideband Hybrid Amplifier," *Technical Digest OFC 2000,* paper TuA1, March 2000.
67. Y. Sun, A.K. Srivastava, J. Zhou, and J.W. Sulhoff, "Optical Fiber Amplifiers for WDM Optical Networks," *Bell Labs Technology Journal, 4,* 1, 187–206, 1999.
68. B. Giles and E. Desurvire, "Modeling Erbium-Doped Fiber Amplifiers," *Journal Lightwave Technol., 9,* 271–283, February 1991.
69. E. Desurvire, *"Erbium-Doped Fiber Amplifiers,"* Wiley, 1994.
70. A. Saleh, R. Jopson, J. Evankov, and J. Aspen, "Modeling of Gain in Erbium-Doped Fiber Amplifiers," *IEEE Photon. Technol. Lett., 2,* 714–717, Oct. 1990.
71. R.J. Mears, L. Reekie, I.M. Jauncey, and D.N. Payne, "Low-Noise Erbium-Doped Fiber Amplifier Operating at 1.54 μm," *Electronics Letters, 23,* 19, 1026–1028, September 1987.
72. E. Desurvire, J.R. Simpson, and P.C. Becker, "High-Gain Erbium-Doped Traveling-Wave Fiber Amplifiers," *Optics Letters, 12,* 11, 888–890, November 1987.
73. M.A. Muriel, J. Azana, and A. Carballar, "Fiber Grating Synthesis by Use of Time-Frequency Representations," *Optics Letters, 23,* 1526–1528, 1998.

74. K. Inoue, "Phase-Mismatching Characteristics of FWM in Fiber Lines with Multi-Stage Optical Amplifiers," *Optics Letters, 17,* 801–803, 1992.
75. K.O Hill, D. C. Johnson, B. S. Kawasaki, and R.I. MacDonald, "CW Three-Wave Mixing in Single-Mode Optical Fibers," *Journal Appl. Phys., 49,* 5098–106, 1978.
76. K. Inoue, "Four-wave Mixing in an Optical Fiber in the Zero-Dispersion Wavelength Region," *Journal Lightwave Tech., 10,* 11, 1553–61, November 1992.
77. C. Palmer, "*Diffraction Grating Handbook,*" www.thermorgl.com April, 2002.
78. S.V. Kartalopoulos, *Next Generation SONET/SDH,* Wiley/IEEE Press, 2004.
79. S.V. Kartalopoulos, *Understanding SONET/SDH and ATM: Communications Networks for the Next Millenium,* IEEE Press, 1999.
80. B. Mukherjee, *Optical Communication Networks,* McGraw-Hill, 1997.
81. T-H Wu, *Fiber Network Service Survivability,* Artech House, 1992.
82. Y. Miyao and H. Saito, "Optimal Design and Evaluation of Survivable WDM Transport Networks," *IEEE Journal on Selected Areas in Communications, 16,* 7, 1190–1198, September 1998.
83. B. Van Caenegen, W. Van Parys, F. De Turck, and P. M. Demeester, "Dimensioning of Survivable WDM Networks," *IEEE Journal on Selected Areas in Communications, 16,* 7, 1146–1157, September 1998.
84. S.V. Kartalopoulos, "A Plateau of Performance?," *IEEE Communications Magazine,* 13–14, September 1992.
85. E. Karasan and E. Ayanoglu, "Performance of WDM Transport Networks," *IEEE Jl on Selected Areas in Communications, 16,* 7, 1081–1096, Sept. 1998.
86. L.G. Raman, *Fundamentals of Telecommunications Network Management,* IEEE Press, 1999.
87. S.V. Kartalopoulos and M. Bouhiyate, "Modeling a Supercontinuum Source for 20-channel WDM Applications," in *Proceedings of the Fifteenth IASTED International Conference on Modeling and Simulation,* Marina Del Rey, CA, March 1–3, 2004, pp. 296–301.
88. B.T. Doshi, S. Dravida, P. Harshavardhana, O. Hauser, and Y. Wang, "Optical Network Design and Restoration," *Bell Labs Technical Journal, 4,* 1, 58–84, 1999.
89. M.A. Marsan, A. Bianco, E. Leonardi, A. Morabito, and F. Neri, "All-Optical WDM Multi-Rings with Differentiated QoS," *IEEE Communications Magazine, vol 37,* 2, 58–66, Feb. 1999.
90. S.V. Kartalopoulos, "A Manhattan Fiber Distributed Data Interface Architecture," in *Proceedings of Globecom'90,* San Diego, December 2–5, 1990, pp. 141–143.
91. S.V. Kartalopoulos, "Disaster Avoidance in the Manhattan Fiber Distributed Data Interface Network," in *Proceedings of Globecom'93,* Houston, TX, December 2, 1993, pp. 680–685.
92. Y. Chen, M.T. Fatehi, H.J. LaRoche, J.Z. Larsen, and B.L. Nelson, "Metro Optical Networking," *Bell Labs Technology Journal, 4,* 1, 163–186, 1999.
93. A.R. Chraplyvy, "High-Capacity Lightwave Transmission Experiments," *Bell Labs Technology Journal, 4,* 1, 230–245, 1999.
94. D.B. Buchholz et al., "Broadband Fiber Access: A Fiber-to-the-Customer Access Architecture," *Bell Labs Technology Journal, 4,* 1, 282–299, 1999.
95. M. Berger et al., "Pan-European Optical Networking Using Wavelength Division Multiplexing," *IEEE Com. Mag., 35,* 4, 82–88, 1997.

96. B. Fabianek, K. Fitchew, S. Myken, and A. Houghton, "Optical Network Research and Development in European Community Programs: From RACE to ACTS," *IEEE Com. Mag., 35,* 4, 50–56, 1997.
97. R.K. Snelling, "Bringing Fiber to the Home," *IEEE Circuits and Devices Magazine, 7,* 1, 23–25, January 1991.
98. A. McGuire, "Architectural Models for Supervisory and Maintenance Aspects of Optical Transport Networks," paper presented at ICC'97 Workshop on WDM Network Management and Control, Montreal, Quebec Canada, June 1997.
99. E. Goldstein, "The Case for Opaque Multiwavelength Lightwave Networks," paper presented at ICC'97 Workshop on WDM Network Management and Control, Montreal, Quebec Canada, June 1997.
100. D.K. Hunter et al., "WASPNET: A Wavelength Switched Packet Network," IEEE Comm. Magazine, 37, 3, 120–129, March 1999.
101. D. Banerjee, J. Frank, and B. Mukherjee, "Pasive Optical Network Architecture Based on Waveguide Grating Routers," *IEEE Journal on Selected Areas in Communications, 16,* 7, 1040–1050, September 1998.
102. S.V. Kartalopoulos, "Fault Detectability: A prerequisite for Protection and Restoration in Next Generation DWDM Networks," paper presented at Globecom-2003 Conference, Workshop in Network Survivability, San Francisco, December 1–4, 2003.
103. N. Takato et al., "128-Channel Polarization-Insensitive Frequency-Selection-Switch Using High-Silica Waveguides on Si," *IEEE Photon. Technol. Lett., 2,* 6, 441–443, 1990.
104. S.V. Kartalopoulos and M. Bouhiyate, "Supercontinuum Generation in Highly Nonlinear Dispersive Fiber," to be presented at IASTED/Modeling and Simulation, Marina Del Rey, CA, March/1–3/2004.
105. S.V. Kartalopoulos, "Consumer Communications in the Next Generation Access Network," in *CCNC-2004 Conference,* Las Vegas, January 5–8, 2004, session N13, paper no. 0-7803-8146-7.
106. M. Bouhiyate and S.V. Kartalopoulos, "Performance of Optimized Chromatic Dispersion Compensated 10 Gbps DWDM Links," in *Proceedings of NFOEC-2003 Conference,* Orlando, FL, September 6–10, 2003, pp. 1162–1167.
107. S.V. Kartalopoulos, "Ultra-High Bandwidth Data Transport for DWDM Short, Medium-Haul and Metro Using Low Bit Rates," in *APOC 2002 Conference,* Optical Switching and Optical Interconnection II, pp. 100–107, October 2002, Shanghai, China.
108. S.V. Kartalopoulos, "Factors Affecting The Signal Quality, and Eye-Diagram Estimation Method for BER and SNR in Optical Data Transmission," in *Proceedings of the International Conference on Information Technology,* April 5–7, 2004, Las Vegas, pp. 615–619.
109. S.V. Kartalopoulos, "Ultra-fast Pattern Recognition in Broadband Communications Systems" in *ISPACS'98,* November 1998, Melbourne, Australia, pp. 558–591.
110. S.V. Kartalopoulos and P. Verma, "Pragmatic Teaching of Advanced Optical Networks: Connecting Physics, Optical technology and Networks," in *Proceedings of the IEEE International Conference on Information Technology: Coding and Computing* (ICTT 2003), pp. 65–68, Las Vegas, 28–30 April 2003.
111. J.J. Sluss, Jr., S.V. Kartalopoulos, H.H. Refai, M.J. Riley, and P.K. Verma, "The Telecommunications Interoperability Laboratory," in *Proceedings of ASEE-2003,* Nashville, TN, 23 June 2003.

STANDARDS

1. ANSI/IEEE 812-1984, *Definition of Terms Relating to Fiber Optics*, 1984.
2. IEC Publication 793-2, Part 2, *Optical Fibres—Part 2: Product Specifications*, 1992.
3. IEC Publication 1280-2-1, "Fibre Optic Communication Subsystem Basic Test Procedures; Part 2: Test Procedures for Digital Systems; Section 1—Receiver Sensitivity and Overload Measurement."
4. IEC Publication 1280-2-2, "Fibre Optic Communication Subsystem Basic Test Procedures; Part 2: Test Procedures for Digital Systems; Section 2—Optical Eye Pattern, Waveform and Extinction Ratio Measurement".
5. ITU-T Recommendation G.650, *Definition and Test Methods for the Relevant Parameters of Single-Mode Fibres*, 1997.
6. ITU-T Recommendation G.661, "Definition and Test Methods for the Relevant Generic Parameters of Optical Fiber Amplifiers," November 1996.
7. ITU-T Recommendation G.662, "Generic characteristics of Optical Fiber Amplifier Devices and Sub-Systems," July 1995.
8. ITU-T Recommendation G.663, "Application Related Aspects of Optical Fiber Amplifier Devices and Sub-Systems," October 1996.
9. ITU-T Draft Recommendation G.664, "General Automatic Power Shut-down Procedure for Optical Transport Systems," October 1998.
10. ITU-T Recommendation G.671, *Transmission Characteristics of Passive Optical Components*, 1996.
11. ITU-T Recommendation G.681, "Functional Characteristics of Interoffice and Long-Haul Line Systems Using Optical Amplifiers, Including Optical Multiplexers," June 1986.
12. ITU-T Draft Rec. G.692, "Optical interfaces for Multi-Channel Systems with Optical Amplifiers," October 1998.
13. ITU-T Recommendation G.702, *Digital Hierarchy Bit Rates*, 1988.
14. ITU-T Recommendation G.707, "Network Node Interface for the Synchronous Digital Hierarchy," 1996.
15. ITU-T Draft Rec. G.709, "Network Node Interface for the Optical Transport Network (OTN)," October 1998.
16. ITU-T Draft Rec. G.798, "Characteristics of Optical Transport Networks (OTN) Equipment Functional Blocks," October 1998.
17. ITU-T Rec. G.805, "Generic Functional Architecture of Transport Networks," October 1998.
18. ITU-T Recommendation G.841, *Types and Characteristics of SDH Network Protection Architectures*, 1996.
19. ITU-T Draft Rec. G.871, Framework for Optical Networking Recommendations," October 1998.
20. ITU-T Recommendation G.872, *Architecture of Optical Transport Networks*, 1999.
21. ITU-T Draft Rec. G.873, "Optical Transport Network Requirements," October 1998.
22. ITU-T Draft Rec. G.874, "Management Aspects of the Optical Transport Network Element," October 1998.
23. ITU-T Draft Rec. G.875, "Optical Transport Network Management Information Model for the Network Element View," October 1998.

24. ITU-T Recommendation G.911, "Parameters and Calculation Methodologies for Reliability and Availability Of Fibre Optic Systems," 1993.
25. ITU-T Recommendation G.957, "Optical Interfaces for Equipments and Systems Relating to the Synchronous Digital Hierarchy," 1995.
26. ITU-T Recommendation G.958, "Digital Line Systems Based on the Synchronous Digital Hierarchy for use on Optical Fibre Cables," 1994.
27. ITU-T Draft Rec. G.959, "Optical Networking Physical Layer Interfaces," February 1999.
28. ITU-T Recommendation L.41, *Maintenance Wavelength on Fibres Carrying Signals,* May 2000.
29. Telcordia (previously Bellcore) GR-253, "Synchronous Optical Network (SONET) Transport Systems: Common Generic Criteria," Issue 2, December 1995.
30. Telcordia (previously Bellcore) GR-1377, "SONET OC-192 Transport Systems Generic Criteria," Issue 3, August 1996.
31. Telcordia (previously Bellcore), TR-NWT-233, "Digital Cross Connect System," November 1992.
32. Telcordia (previously Bellcore), TR-NWT-499, *Transport Systems Generic Requirements (TSGR): Common Requirements,* Issue 5, December 1993.
33. W. Simpson, "PPP over SONET/SDH," IETF RFC 1619, May 1994.
34. Telcordia (previously Bellcore), TR-NWT-917, "Regenerator," October 1990.

CHAPTER 5

Noise Sources Affecting the Optical Signal

5.1 INTRODUCTION

This chapter attempts to provide a simplified interpretation of the meaning of "noise," present its underlying theories, and enumerate various noise sources that contaminate the optical signal. In subsequent chapters, we examine the manifestation of noise, how it degrades the performance of the optical receiver, and how it is detected.

5.1.1 What Is Noise?

Talking about noise, the first thing that comes to mind is acoustical noise. Perhaps this is because acoustical noise is ubiquitous and we experience it everyday. In fact, although we live immersed in noise, the auditory ear the ability to distinguish a signal even if it is severely contaminated with other sources that are classified as noise. In many cases, it has the remarkable ability to adapt to noise conditions within a short period and minimize or ignore noise. This adaptability becomes evident when one enters an anechoic chamber; our adaptation to noise is so formidable that now the absence of noise is annoying!

Thus, the definition of noise is subjective and it can be as simple as "the superposition of a time-varying undesired signal on another signal," or as complex as "the mean squared fluctuation of photon phase per unit bandwidth caused by Poissonian or Gaussian . . . ," and so on; we shall see that, depending on specific conditions, the definition of noise varies.

In communications, noise corrupts the integrity of signal, that is, the true value of bits in the signal, and under certain conditions bits are so badly degraded that they cannot be clearly distinguished by the photodetector.

5.1.2 Sources of Noise

The study of noise in communications started very early in the last century with Albert Einstein, who in 1909 realized that the electromagnetic fluctuation between particles and waves is different, and with Brown, who studied the (statistical) random movement of electrons in a conductor. In 1928 two papers on electron thermal noise were published back to back and in the same issue of *Physical Review,* one authored by J. B. Johnson, setting the theoretical foundation, and the other authored by H. Nyquist explaining its applicability. Schottky, however, had already explained in 1918 the statistical movement of electrons in a vaccum thermionic tube. At that time, the study of noise was more targeted to electron movement than photon movement. However, many of the concepts that were developed are currently applicable to the study of photon noise either because of the statistical movement of photons or because of their quantum-mechanical generation. Thus, the study of noise started with examining the movement of electrons (as charged particles) in conductors, electronic tubes, and semiconductors, and of electromagnetic waves in space or in dielectric waveguides.

Noise is omnipresent in any conductor, even in the absence of external voltage across it. Thus, when a signal travels in a conductor, is it the noise that contaminates the signal, or is it the signal that contaminates the noise? Refering to this, Landauer used to say, "there is signal in the noise."

As noise became better understood and the generating mechanisms were identified, it began to be classified based on origin, generating mechanism, and effect that it had on the propagating signal.

Based on the mechanisms by which it is induced in the signal, noise may classified as *extrinsic* or *intrinsic*:

- Extrinsic noise is unwanted frequencies superimposed on the signal from external sources, such as electromagnetic interference, amplifier noise (ASE), cross talk, four-wave mixing, or detector noise.
- Intrinsic noise is unwanted random energy fluctuation of the signal electrons or photons caused by temperature, by the nonlinearity and gradient nonlinearity of the transmission medium, or by contaminants in the matrix of the medium; such noise is thermal noise, shot noise, and flicker or 1/f noise.

Based on the effect that noise has on the signal, the most common types of noise are:

- White (random) frequency modulation (WFM)
- White (random) phase modulation (WPM)
- Flicker phase modulation (FPM)
- Flicker frequency modulation (FFM)
- Random walk FM (RWFM)

The bottom line in the signal performance of communications is not whether there is noise or not, but how much noise power has been added to the power of the signal. Thus, one metric is the *signal-to-noise ratio* (SNR). If this ratio refers to optical noise and signal, then it is called OSNR. Thus, OSNR provides a measure of the clarity (that also includes intereferences) of the transmitted optical signal when it arrives at the photodetector. Unfortunately, even if OSNR arrives with the expected performance, the receiver itself is a source of noise, such as thermal, shot, flicker, amplifier, and others, as discussed in Chapter 3.

In addition, there are other noise sources that cause fluctuation of the sampled signal, such as oscillator noise. This, too, is added to the electrical signal (after been converted from the optical signal) at the receiver.

5.1.3 Oscillator Noise

When an optical signal arrives at the photodetector, it is already contaminated with noise and degraded (shape distortion, attenuation). The photodetector detects photons and generates electrons, which ideally are proportional to photons received. However, the photodetector is an electronic device that itself is a source of noise. The electrical signal needs to be sampled by a clock at a specific periodic instance, the point of which has been determined such that it remains immune to jitter and minimizes the bit error rate (see Chapter 6). However, sampling is as good as the accuracy and stability of the clock. If the clock frequency fluctuates, this fluctuation is manifested as oscillator noise, which further degrades the received signal. A treatment of this noise has been provided by ITU-T (G.812), which defines it as the amount of phase noise generated by the clock in the holdover state or by a slave clock and produced at the output when there is an ideal input reference signal. As a consequence, a frequency stability criterion of the reference clock is recommended to be 10 times better than the output requirement. In addition to frequency stability, noise tolerance sets the lower limit of the maximum phase noise level at the input of the clock that should be accommodated while:

- The clock is maintained within prescribed performance limits.
- The clock does not trigger any alarms.
- The clock remains on the same reference source (it does not select another reference).
- The clock remains in lock state (it does not go into holdover).

5.1.4 Statistical Noise

Noise generation is studied using the statistical behaviour of the generating mechanism. Since noise is the result of statistical movement of charged carriers or photons, noise is classified according to its statistical distribution as Gaussian, Poisson, Fermi, Bose–Einstein, and so on.

Thermal (or Johnson) noise is the quantum-statistical phenomenon of random movement of charged carriers (such as electrons) due to thermal agitation. Its initial study was based on a classical model of passive devices (R, C, L).

Shot noise is the random and time-dependent fluctuation in discrete-electron flow, as opposed to a steady flow of electrons in a conductor. That is, each electron is independent from the other electrons in the flow. In photon propagation, it is due to the random fluctuation of photon phase.

Impurities in the conductor region of semiconductor MOSFET devices trap charged carriers. In these traps, the carriers remain for a short but random time. When carriers are released, they generate a time-dependent current fluctuation that is manifested as noise. Because of the time dependency, which is characteristic at frequencies below 10 kHz, this type of noise is called *flicker* or 1/f noise.

Amplifier spontaneous emission (ASE) is the random emission of photons by excited atoms in the fiber, or in an excited semiconductor photonic device. These random photons are in the same spectral band as the signal and are manifested as noise.

Cross talk, dispersion, and dispersion slope, PMD, PDL, FWM, XPM, MI, ISI, and so on, are also noise sources due to the nonlinear behavior of fiber in conjunction with the signal strength and bit rate.

In the following, we look into certain noise sources in greater depth.

5.2 TRANSMISSION FACTORS

As a signal propagates, its power is attenuated; the longer the distance, the more the attenuation. However, power attenuation is not the same for all frequencies as it is a function of signal frequency.

Loss affects the receiver's ability to reconstruct the transmitted signal at the source faithfully, that is, without errors. This is closely related to the receiver's sensitivity level.

As an optical signal departs from the laser and propagates in the fiber, it is contaminated with noise and jitter. Several mechanisms, external or internal, are responsible for the addition of noise, including the transmitter itself and the receiver. Although noise is undesirable, what is more important is the power of noise in relation to the power of the signal, measured as the signal to noise ratio (SNR). The amount of noise in the signal, and thus the SNR, is closely related to the discrimination ability of the receiver; that is, its ability to separate the signal from the signal plus noise. It has been established that at the receiver, an acceptable SNR value over the path is 40 dB.

Signal power loss, jitter, and SNR degradation are manifested at the receiver as:

- An analog signal is distorted so that it cannot always be reproduced faithfully, resulting in loss of information (Figure 5.1).
- A digital signal is distorted so that the value of the original bit symbols cannot always be recovered, increasing the bit error rate, thus resulting in loss of information (Figure 5.1)

5.2.1 Phase Distortion

WDM signals propagate over the fiber transmission medium with a dielectric constant that depends on optical power and wavelength. Thus, as each wavelength

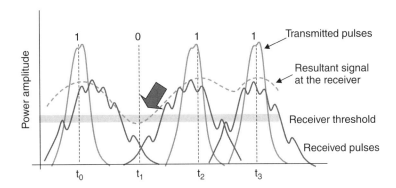

Figure 5.1. Compound effect of loss, noise, and jitter on SNR and BER on digital signals. The arrow shows the area for possible bit error at the receiver.

propagates at a different speed, the phase relationship of the WDM signals does not remain uniform. This is manifested as signal-phase distortion. To minimize phase distortion, the characteristics of the medium are manipulated to the degree possible so that the operating spectral band is an almost zero differential delay region (Figure 5.2).

The degradation of the quality of signal is also manifested by the eye diagram of the signal. The eye diagram is a visual method using a testing apparatus (Figure 5.3). In subsequent chapters, we shall see that the eye-diagram method can become powerful in estimating signal performance parameters.

5.2.2 Frequency Distortion

The monochromaticity of laser light depends on the uniformity and stability of the active region of the laser device. However, many random and periodic fluctuations influence the device, which affects the modulated light in both spectral content and timing, such as chirp. In addition, photonic pulses are affected by fiber nonlinearities and nonuniformities that have similar noise effects.

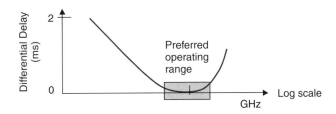

Figure 5.2. The preferred operating spectral band is in the vicinity of zero differential delay.

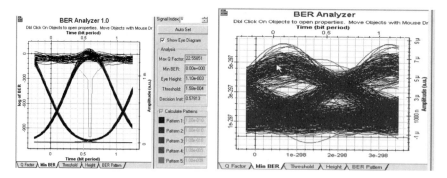

Figure 5.3. The eye diagram provides a qualitative visual of the quality of the received signal. Left, a good quality signal (open eye); right, a bad quality signal (closed eye).

5.2.3 Polarization Distortion

Photon polarization is distorted by mechanisms similar to those that cause frequency distortions. Parameter fluctuations in laser and other nonlinear components, including fiber, cause fluctuations in polarization.

5.2.4 Echo

As the signal propagates, a small part of its power is reflected by discontinuities or mismatches on the optical path. When the reflection reaches the optical source after some *round-trip delay* Δt it is mixed with the transmitted signal (at the transmitter) and, depending on the power of echo at the transmitter and the phase relationship with the transmitted signal, it may have a deleterious effect. Here, we assume that the interaction between the propagating signal in one direction and echo in the opposite direction is negligible.

Two parameters related to echo are of importance: echo coefficient (EC) and echo return loss (ERL).

Echo coefficient is defined as the ratio of incident signal power to the reflected power by the discontinuity

$$EC = II/IR$$

measured at the discontinuity. However, if the reflected power is very small and the distance between transmitter and discontinuity is long, then the echo signal may be insignificant due to attenuation. If the expected echo is significant, then a directional isolator on the optical path minimizes echo. In this case, the insertion loss of the isolator should be accounted for in link power calculations.

Echo return loss is defined as the ratio of reflected power to incident power at the discontinuity:

$$ERL = IR/II$$

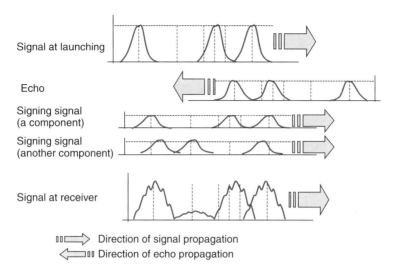

Figure 5.4. Effect of singing on original signal (oversimplified and exaggerated to illustrate the effect).

measured at the discontinuity, where IR and II are the reflected intensity and incident intensity of the signal, respectively.

5.2.5 Singing

Under certain conditions, the echo signal reverberates between two discontinuities. This is known as singing. Since the signing component has the same wavelength as that of the original signal, any acquired impairments, such as noise and jitter, will be added to the signal. In addition, since reverberating components are out of phase with the original signal, they, too, will contribute to the original signal's dispersion (Figure 5.4).

To minimize echo and singing, a directional isolator is inserted in the optical path. However, an isolator has insertion loss and it should be taken into account in the power budget of the link.

5.3 THERMAL NOISE

The Brownian (random) motion of electrons agitated due to temperature generates in a conductor a random current that is termed *thermal noise*. The study of thermal noise starts with the assumption that n electrons in a conductor of length l and cross-sectional area A behave like a charged "gas" in which each electron moves due to heat agitation, like any particle does in a gas. Each electron in the "gas" moves in all possible directions. In their random movement, electrons collide and change direction. If the electron density is n and the mean time between collisions is

τ (in metals $\tau < 10^{-13}$ sec) then, within a period of observation t, electrons experience on the average t/τ collisions. Similarly, the total number of electrons in the "gas" is nAl, the resistance is $R = 2ml/(ne_0^2 \tau A)$, where m is the electron mass, and the average kinetic energy of each electron is (Figure 5.5)

$$E_{\text{kin, avg}} = \tfrac{1}{2} kT$$

in joules, where T is the temperature in degrees Kelvin and k is the Boltzmann's constant (1.38×10^{-23} J/K or Ws/K).

As the ith electron moves, it generates a current equal to its charge divided by the mean time collision τ, $I_i = e/\tau$. Such current, if it exists, is called *thermal noise current*. Since the electron collides with other electrons, it changes direction after each collision and, thus, its current changes direction. Thus, the net thermal noise current over a long period t is the vectorial sum of all τs, which for truly random movements is zero. Thus, for one electron or for n electrons in this "gas," the net thermal noise current is zero, or, statistically speaking, the expected mean value of the thermal noise current is zero or $E(I_{ThN}) = 0$. However, the expected value of I_{ThN}^2 in time t is not zero since by squaring, negative vectors become positive. In this case,

$$E(I_{ThN}^2) = (e^2/l^2) nAl \tau^2 kT/m\tau t$$
$$= ne^2 \tau A 2kT/(2mlt)$$
$$= 2kT/(Rt)$$

When a voltage V is applied across the resistor, the random movement of the electron "gas" changes since now the directional movement is the result of thermal agitation

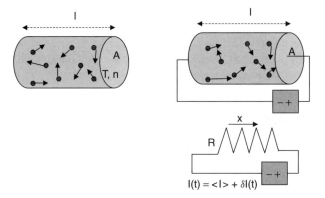

Figure 5.5. Left: Model of a resistor with "gas" of electrons, each electron moving in a random direction due to heat. Right: Electrons moving in a biased resistor because of heat and drift.

and the electric field V/L that causes them to drift in the direction of the field (in the x-direction and toward the positive pole). Thus, at any instant there are two velocity vectors (drift and thermal), the sum of which constitutes the net velocity of electrons:

$$U_x = U_{\text{Drift}} + U_{Th}$$

These velocities correspond to a net current that consists of the mean drift and mean thermal currents (Figure 5.6):

$$I(t)_x = \langle I_{\text{Drift}} \rangle + \delta I_{Th}(t)$$

Now, the work required by N electrons to cross the resistor is NeV and the change of work required to change the random movement of electron in the "gas" in the direction x is $U_{Th}\tau eV/L$. These two quantities must be equal at equilibrium and, thus, $NeV = U_{Th}\tau eV/L$. The number of electrons in the x-direction generates a net charge $Q = Ne$, the current of which as it passes through a resistor is, according to Ohm's law, $V = IR$. Thus, the net charge that the applied voltage must cause to flow is $Q = Ne = U_{Th}\tau e/L$.

To study the thermal noise generated in a transmission line, we assume a model that consists of two load resistors, $R1$ and $R2$ interconnected with a pair of lossless (non-resistive) transmission lines of length L (Figure 5.7), and a signal at frequency f propagated over it at a velocity u.

A transmission line is capable of propagating a signal in two directions or modes. The energy of the signal in one of the directions is

$$E = h\nu/[e^{h\nu/kT} - 1]p(f)$$

Where h is Planck's constant, k is Boltzmann's constant, and $p(f)$ is the thermal average energy.

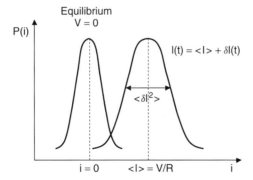

Figure 5.6. The current generated at equilibrium (no voltage) has a zero mean value. However, when a voltage is applied, then there is a drift and thermal noise.

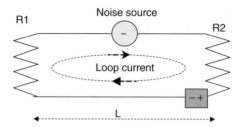

Figure 5.7. Model of a lossless transmission line.

In the classical limit, $h\nu \ll kT$ and the thermal energy per direction may be assumed to be kT. Thus, the energy on the transmission line and within the frequency band Δf is

$$E(\Delta f) = 2kT\Delta f(L/u)$$

The quantity $1/d\tau = u/L$ is known as Planck's factor. Thus, the rate of energy (or power) that flows at one end of the transmission line (that is, in one mode at L) is

$$P(\Delta f) = kT\Delta f$$

And

$$P(\Delta f) = \langle I^2 \rangle R$$

where $\langle . \rangle$ is the mean. Similarly, the mean square voltage developed across the resistor $R2$ in the two-resistor ($R1$ and $R2$) network is

$$\langle I^2 \rangle R = \langle V^2 \rangle R2/(R1 + R2)^2$$

Now, assuming that the loads are matched, $R = R1 = R2$, then $\langle V^2 \rangle = 4\langle I^2 \rangle R$ and, thus

$$\langle V^2 \rangle = 4kTR\Delta f$$

The latter is the Nyquist theorem that expresses the mean square voltage (or the variance of voltage) in thermal equilibrium at temperature T across a load resistance and for a bandwidth Δf. Similarly, the mean squared noise current per frequency interval is $\langle I^2 \rangle = 4kT/R$.

From the Nyquist theorem, the thermal noise power is obtained as

$$N_{ThN} = 4kT\Delta f$$

Where Δf is the observing spectral bandwidth (or the bandwidth of the filter at the receiver). It is noticeable that thermal noise is independent of frequency and load

5.3 THERMAL NOISE

resistance. Based on the Nyquist theorem, the spectral density $G(f)$ is the mean-squared current fluctuation per unit bandwidth:

$$G(f) = \langle dI(f)^2 \rangle / \Delta f$$

At equilibrium, the latter becomes *independent of frequency*:

$$G(f) = 4RkT$$

From this, the maximum thermal noise power per unit frequency range delivered by a resistor to a matching load R is

$$G(f)/R = 4kT$$

Assuming that the voltage fluctuation is Gaussian, then the mean noise voltage across the load resistor is

$$\langle V \rangle = 0$$

and the standard deviation is

$$\sigma(V) = \sqrt{\langle V^2 \rangle} = 1.3 \times 10 - 10 \, (R\Delta f)^{1/2} \qquad \text{volts at room temperature } (300°K)$$

The Nyquist theorem may also be expanded to include the electrical conductivity $1/\rho$, which is defined from

$$1/\rho = Ne^2 \tau / m$$

and

$$R = \rho L / A$$

The conductivity ρ is derived from the drift velocity equation:

$$du/dt = eE/m - u/\tau$$

the solution of which at steady state is

$$u = eE\tau/m$$

From this, the drift velocity per unit electric field or *mobility* is

$$\mu = u/E = e\tau/m$$

and the electrical conductivity σ is

$$\sigma = 1/\rho = j/E = Neu/E = Ne^2\tau/m$$

where j is the current density.

5.3.1 Thermal Noise at the Receiver

The random motion of electrons takes place also in the receiver itself. As the generated current in it flows through a load resistor, the mean value of the thermal noise current, the current standard deviation I^2_{ThN}, is expressed as

$$I^2_{ThN} = 4kTB/R \qquad (pA^2)$$

where T is the temperature (in degrees Kelvin), B is the bandwidth or frequency range of the passband filter at the photodetector, k is Boltzmann's constant equal to 1.38×10^{-23} J/K (or watt × second/K), and R is the load resistance. In practice, B varies from $\frac{1}{2}T$ to $1/T$, where T is the bit period.

The one-sided power spectral density of thermal noise (for a nonband-limited system) is $P^+_{ThN} = kT$ (for 1 Hz bandwidth). Thus, the thermal noise power P_{ThN} (per 1 Hz bandwidth) and at absolute zero is:

$$P_{ThN} = 4.00 \times 10^{-21}\ W \times s/Hz = -204\ dB = -174\ dBm$$

The *noise figure*, *NF*, is a measure of the amount of noise added to the signal. If the effective noise temperature of a device is T_e, then, the *NF* (in dB) is

$$NF = 10\ \log(1 + T_e/290)$$

If the signal power is expressed in terms of the photocurrent and the load resistor, then the signal to noise ratio (SNR) is expressed in terms of the combined shot noise and thermal noise current variances:

$$SNR = 10\ \log\ [\text{signal power}/(\text{noise power})] = 10\ \log\ [I^2_{ThN}/(I^2_{SN} + I^2_{ThN})] \qquad (dB)$$

The latter relationship assumes an incoming signal free from noise. In reality, the incoming photonic signal already contains optical noise generated by several sources. As a result, the optical noise needs to also be included in the total noise power in order to calculate a realistic signal-to-noise ratio:

$$SNR = 10\ \log\ [\text{signal power}/(\text{total noise power})] = 10\ \log\ [I^2_{ThN}/(\Sigma(I^2_{\text{Noise-j}}))] \qquad (dB)$$

where $I_{\text{Noise-j}}$ indicates the current generated by any source but the signal.

5.4 SHOT NOISE

With the deployment of thermionic tubes, another type of noise was noticed that could not be explained with the prevailing noise theories. Schottky termed this *shot noise*. At a later time, it was discovered that shot noise is also generated in solid-state devices such as Schottky diodes and p-n junction diodes. Schottky studied the emission of overheated electrons from the cathode of thermionic tubes, which, once released from the cathode, are accelerated by the electric field and hit the anode randomly. The term "shot" was an analog of a shotgun's pellets "hitting" a target. Interestingly, Schottky's analog is closer to photonic emission than electronic emission; an analogy more appropriate to thermionic emission of electrons would be the droplets of a shower head hitting the floor of the tub as these droplets have been accelerated by the gravitational field (Figure 5.8).

5.4.1 Electronic Shot Noise

In a thermionic tube, the speed of released electrons from the cathode is Poissonian with mean velocity $\langle U_{C,i} \rangle$, mean energy $\langle E_{S-N} \rangle$, and mean total charge $\langle Q \rangle = \langle Ni \rangle e$, where $\langle Ni \rangle$ is the mean number of electrons emitted from the cathode. As soon as electrons are emitted, a "cloud" of electrons exists temporarily just above the cathode. These electrons repel the electrons that are just emitted from the cathode, causing them to slow down or change direction away from the anode. However, as the anode pulls the electrons towards it, the cloud looses its repulsion capability and allows new electrons to start moving away from the cathode, which in turn form another cloud, and so on. The electrons that move toward the anode move independently from each other and arrive at the anode with random velocity and mean velocity $\langle UA,i \rangle$, and they generate a fluctuating DC current, $I(t) = I + dI$ (measured in number of charges per second), where I is the DC component and dI is the fluctuation. The corresponding spectral density, including both positive and negative fluctuations, is $G = 2e\langle I \rangle$.

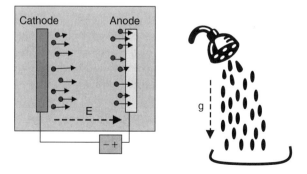

Figure 5.8. An analogy for shot noise (right) of a thermionic tube (left).

Since electrons arrive randomly at the anode, each electron generates a current $I(t)$ during the transit time τ from cathode to anode. The pulse generated by each electron may be approximated by a Dirac pulse, whereas the fluctuating current causes a mean-squared shot noise current I_S^2. If the transit time is very small, ~10^{-9} s, then $1/\tau \sim 10^9$ Hz and the spectral density is almost constant over the bandwidth $B = \Delta f$. Then the total energy W in Δf is

$$W = \int_{-\infty}^{\infty} |F(\omega)|^2 df = 2\int_0^{\infty} |F(\omega)|^2 df$$

In terms of bandwidth, B, it is

$$W = 2e^2 B$$

If the average number of n electrons in the interval τ is i_T, and the average shot noise current is I_{SN}, the deviation is

$$\langle i_T^2 \rangle = e^2 \langle n - \langle n \rangle \rangle^2 / \tau^2 = eI_{SN}/\tau$$

For the average number of electrons arriving at the anode within an observation interval T, where $T \gg \tau$, then the average squared shot noise current in a 1 Ω load becomes

$$I_{SN}^2 = nW/T = 2ne^2B/T$$

Substituting for the average current at the anode $I_A = ne/T$, then, the shot noise power is

$$P_{SN} = 2eI_A B$$

This is the Schottky formula for the thermionic shot noise power. Because of certain assumptions made in this derivation, Schottky's formula is correct for very high frequencies.

In short, shot noise is the random and time-dependent fluctuation in discrete-electron flow, as opposed to a steady flow of electrons in a conductor. In photon propagation, it is the random fluctuation of the photon phase. Based on the aforementioned definitions, the shot noise current standard deviation, I_{SN}^2, and average power P_{SN} in the bandwidth B (in MHz) and at a load R at the photodetector are provided by the relationships

$$I_{SN}^2 = 2e(I_{\text{dark}} + I_{ph})B$$

and

$$P_{SN} = 2e(I_{\text{dark}} + I_{ph})RB$$

where I_{dark} is the current that flows at the absence of the photonic signal, hence *dark current*, I_{ph} is the photocurrent generated by the photonic signal, R is a resistive load, and B is the bandwidth of the photodetector (in MHz).

Dark current is the current that flows in the absence of the photonic signal and photocurrent is that which is generated by the photonic signal.

Thus, for the one-sided power-spectral density, the random noise is

$$S^+ = e\,(I_{dark} + I_{ph})R$$

5.4.2 Shot Noise and the Dual Nature of Photons—The Fano Factor

Lasers generate (almost monochromatic and coherent) photons. However, there is still shot noise, which now is not manifested as a fluctuation of current but as a random fluctuation of phase among photons. Shot noise treats the optical signal as a classical electromagnetic wave. However, the estimation of photodetector sensitivity uses the quantum-mechanical model. With this model, the photodetector responds to the random illumination of photons in the photosensitive region and it generates carriers, creating a random fluctuation of current. The random fluctuation within a period emulates Poisson statistics.

With respect to the particle–wave duality of photons, noise is generated by particles or waves (as defined by the classical theory), and a significant statistical difference exists between them. It has been established (theoretically and experimentally) that in waves energy (noise) fluctuates linearly with the mean energy. In contrast, in particles the energy (noise) fluctuates as with square root of the mean energy. In fact, identifying the noise relationship with respect to mean energy, one can deduce whether the "photon" is more a wave or a particle. To go a step further, a measure of the statistical behavior of noise power with respect to Poissonian noise power of uncorrelated current pulses can provide this. This is expressed by the ratio "noise power to the Poissonian noise power," $F = P/P_{Poisson}$, which is known as the *Fano factor* (after Ugo Fano who in 1947 studied statistics of ionized particles).

Without getting deep into the subject, the Pauli exclusion principle states that two or more electrons cannot be at the same energy level and with the same spin (that is, they cannot be at the same state). In thermal equilibrium (and not in a conductor), this principle make the Fermi and Boltzmann's statistics indistinguishable. Now, the fluctuation of electrons with energy above eV (the Fermi level) in period τ is $\tau eV/h$. The fluctuation of charge Q follows the binomial statistics with mean-square fluctuation:

$$\langle \delta Q^2 \rangle = e^2[\tau eV]/h]T_i(1 - T_i)$$

Replacing the latter in $P = (2/\tau)\langle \delta Q^2 \rangle$ for a single channel, and summing over n channels, one obtains the general spectral density of noise:

$$P = 2(2e^2/h)eV\Sigma(T_n(1 - T_n)$$

where the summation is from $n = 1$ to N. The latter describes the noise reduction by the factor $(1 - T_n)$ due to Pauli's principle.

Now, the mean Poisonian power noise is $P_{\text{Poisson}} = 2e\langle I\rangle$ and the general power noise is $P = [\langle \delta I(f)^2\rangle/\Delta f] = 2q\langle I\rangle$ (which at equilibrium is $P = 4kT\Delta f$). Then, the Fano factor is expressed as

$$F = 2q\langle I\rangle/2e\langle I\rangle = q/e$$

The latter is a measure of the unit transferred charge; for example, in superconductors, charge is added in Cooper pairs and $q = 2e$ (and thus $F = 2$), whereas in Hall effect conductors (after Robert Laughlin), charge is tunneled in units of a fraction, $q = e/(2p + 1)$, where p is an integer that is determined by the filling fraction $p/(2p + 1)$ of the lowest Landau level. In conductors with long length as compared to the size of atoms, it turns out that the term $(1 - T_n)$ does not exist, $q = e$ and, thus, $F = 1$, indicating that there is no shot noise but only thermal noise.

5.4.3 Shot Noise—Conclusion

Verified by theory and experiment, photons of higher energies (or frequencies) behave more like particles than photons at lower energies, which behave more like electromagnetic waves.

Statistically, coherent radiation follows a Poisson distribution and, thus, $P = P_{\text{Poisson}}$ ($F = 1$). However, as the radiation flows in a nonuniform medium, and if the medium has loss or gain, then $F > 1$.

Thus, shot noise distinguishes deterministic scattering. Particle dynamics is deterministic; the trajectory is determined by the initial position and the momentum vector of a particle.

Wave mechanics is stochastic. The quantum position and momentum are described by Heisenberg's uncertainty principle, which states that the product of the uncertainty of the two is equal to 1. That is, the better the position is known, the more uncertain the moment is, or vice versa. This uncertainty introduces a probabilistic noise. Thus, stochastic scattering is characteristic of waves.

Photons are both particles and waves. Therefore, there must be certain conditions under which a photon behaves more like a particle, and thus is scattered more deterministically (as in reflection and refraction), or more like a wave, and thus is scattered more stochastically (as in diffraction).

A similar uncertainty exists between the number of photons and the phase of photon–waves. If σ_N is the standard deviation of photon number and σ_Φ the standard deviation of photon phase, then $\sigma_N \sigma_\Phi = 1$.

5.5 FLICKER OR 1/f NOISE

Flicker or 1/f noise has been noticed in semiconductor devices with a characteristic frequency dependency. At frequencies below 10 kHz, noise is dominated by time-dependent fluctuations that arise from impurities and is termed flicker or 1/f noise. At high frequencies, noise becomes frequency independent and is termed shot noise.

1/f noise is present in semiconductor devices and it is dominant in metal–oxide semiconductors (MOSs), field effect transistors (FETs), and dielectrics. Specifically, in MOSFET devices, traps in the oxide interface capture electrons or holes. The time constant of these traps has a random or Poisson distribution, and the electrons and holes are released by the trap based on the lifetime or time constant of the trap, which is proportional to $e^{-t/\tau}$. It is these traps and their time constants that respond to low frequency and, thus, 1/f noise is frequency dependent. Now, one approach to evaluate 1/f noise assumes shot noise with a decaying impulse response function; this decay rate is $t^{-1/2}$. Based on this model, 1/f noise is expressed as an infinite sum:

$$F(t) = \Sigma(h(t - t_j))$$

summed from $j = -$ infinity to $j = +$ infinity, where the function $h(t) = Kt^{-1/2}$. In the general case, the exponent can be assumed in the range from 0 to B such as $h(t) = Kt^{-\beta}$ where $0 < t < B$, $0 < \beta < 1$, and K, B, and β are deterministic parameters.

The power spectral density of 1/f noise is then

$$P(f) = \langle I \rangle^2 \delta(f) + \mu \langle K^2 \rangle \Gamma^2(1/2)(2\pi f)^{-1}$$

Where $\delta(f)$ is the dirac function, μ is the rate of the Poisson process, and Γ is the gamma function.

Consolidating several constants in one, A_f, the power spectral density of 1/f noise is finally expressed in a general case as

$$P(f) = A_f/f^b$$

where b is typically about 1. In a MOSFET, $A_f = K_f/[WL\,(C_{oxide})^2]$, where K, W, C_{oxide}, and L are parameters specific to the device.

If all time constants are assumed to be equal, then the latter is simplified to

$$P(f) = \langle I_n^2 \rangle = A_f/f$$

As a consequence, if in a transistor over the bandwidth Δf the thermal noise is $4kT(2/3)g_m\Delta f$ and the flicker noise if $(A/f)\Delta f$, then the total noise power is $\langle I_n^2 \rangle = [4kT(2/3)g_m) + A/f]\Delta f$.

As another consequence, the 1/f spectral density varies quadratically with the mean current I. At high frequency, however, the spectral density becomes linearly proportional to current and it also becomes frequency independent, both of which are the characteristics of shot noise. This is one of the reasons that some view 1/f noise as a type of shot noise that embraces noise types that do not fit in any other category.

5.6 OTHER NOISE SOURCES

5.6.1 Amplifier Noise

In Chapter 4, we discussed several noise sources. Here, we briefly review some of these noise sources. For example, optical amplifiers do not have a flat gain over the amplification spectral band and the deviation from it results in a signal power with a standard deviation σ measured at the output of the filter. If M amplifiers are cascaded in the path, the total standard deviation, or variance σ_{tot}^2 is

$$\sigma_{tot}^2 = \Sigma \sigma_k^2$$

for $k = 1$ to M. Optical amplifiers emit spontaneous noise (ASE).

The EDFA adds amplified spontaneous emission (ASE) noise to the signal, the strength of which is approximated as

$$N_{ASE} = \alpha G_{EDFA} h \nu B$$

where α is a constant (between 1 and 2), G_{EDFA} is the EDFA average gain, h is Planck's constant, ν is the center frequency, and B is the bandwidth. If N_{ASE} represents ASE in one polarization state at the EDFA output, then the noise figure for the EDFA is

$$NF_{ASE} = [(1/G_{EDFA})\{(2N_{ASE}/h\nu B) + 1\}]$$

Substituting N_{ASE} in the latter yields

$$NF_{ASE} = (2\alpha G_{EDFA} + 1)/G_{EDFA}$$

In dB units this is $10 \log_{10} NF_{ASE}$.

Since noise is cumulative, the noise figure over a path with N EDFAs is the sum

$$NF_{N,Total} = 10 \log_{10}[NF_1 + (NF_2 - 1)/G_1 + (NF_2 - 1)/G_1 G_2 \\ + (NF_3 - 1)/G_1 G_2 G_3 + \ldots] \text{ (dB)}$$

The SNR at the receiver, based on ASE noise, is approximated to

$$SNR_{rec} \text{ (dB)} = 58 \text{ (dB)} + P_{in} \text{ (dBm)} - NF_{N,Total} \text{ (dB)}$$

The noise figure (NF) is an indication of the amount of noise added to the signal by the amplification mechanism and it is the dimensionless ratio of input SNR to output SNR of an amplifier:

$$NF = SNR_{in}/SNR_{out}$$

The unsaturated Raman gain is expressed as

$$G_R = I_{out}/I_{in} = \exp[RPL_{eff}/A_{eff}] \exp[-\alpha_s L]$$

Because the Raman gain is not flat, Raman amplification induces a signal power tilt or a slope: a power increase in the shorter-wavelength signals within its amplification range and the opposite in the longer-wavelength signals. The SRS-induced overall power tilt, ΔP, is expressed by:

$$\Delta P = -10 \log[1 - \eta_{pol}\{G_R \Delta f P_{ch} L_{eff} N(N-1)/2A_{eff}\Delta f_R\}]$$

where h_{pol} is a factor of copolarization of the DWDM channels between 1 and 2 (for fully copolarized it is 2 and for noncopolarized it is 1), Δf is the channel spacing, P_{ch} is the power per channel, and N is the number of channels. To amplify a wide spectral range, several Raman pumps are required, the combined gain of which generates not only a tilt but also a gain ripple (see Chapter 4, Figure 4.5).

The noise transfer from the Raman pump to the signal for a copropagating pump is

$$R_s \sim R_P + 20 \log (\ln G_R) + 10 \log[(v_s/L_{eff})^2/\{(\alpha_P v_s)^2 + (2\pi D\Delta\lambda \cdot v_s f)^2\}]$$

and for a counterpropagating pump it is

$$R_s \sim R_P + 20 \log(\ln G_R) + 10 \log[(v_s/L_{eff})^2/\{(\alpha_s v_s)^2 + (2\pi f)^2\}]$$

where the parameters α_P and α_s are the attenuation coefficients at the pump and signal wavelengths, v_s is the velocity of the signal in the fiber, D is the chromatic dispersion of the signal, and $\Delta\lambda$ is the wavelength difference between pump and signal.

The noise figure is a measure of added noise to the signal. Thus, an equivalent (lumped) noise figure, NF_R, is defined as

$$NF_R = [(R_{ASE}/h\nu) + 1]/G_R$$

where R_{ASE} is the Raman ASE density at the end of the fiber, G_R is the on/off Raman gain, h is Planck's constant, and ν is the optical frequency. Similarly, if the fiber span loss is L_f, then the noise figure for the span (without pump) is defined as

$$NF_{unpumped} = 1/L_f$$

Then, the noise figure for the Raman pumped fiber span is

$$NF_R = [P_{noise}/(h\nu\Delta\nu\, G_R L_f)] + 1/(G_R L_f) \quad (dB)$$

where P_{noise} is the noise power in the electrical filter bandwidth (here we assume that the signal has been received and converted to electrical), and G_R is the Raman gain.

Based on the two noise figures, an improvement noise figure is calculated from

$$NF_{improvement} = NF_R - NF_{unpumped} \text{ (dB)}$$

In addition, a penalty in Q-factor is defined, expressed as

$$\text{penalty (dB)} = 10 \log \sqrt{1 + Q^2 \int R_N(f) df}$$

where the integral is from f_1 to f_2 and R_N is the Raman noise transfer from pump to signal (in dB/Hz).

5.6.2 Four-Wave Mixing

Four-wave mixing (FWM), or four-photon mixing under certain conditions, may become a serious source of noise (see Chapter 2). The generated FWM component is

$$E_{FWM} = j[(2\pi\omega)/nc] d\chi \, E_1 E_2 E_3 e^{-\alpha(L/2)} F(\alpha, L, \Delta\beta)$$

where $F(\alpha, L, \Delta\beta)$ is a function of fiber loss, fiber length, and phase mismatch related to channel spacing and dispersion. Since the FWM component with frequency ω is the result of uncorrelated data, it is manifested as random noise superimposed on the channel with linewidth that includes the frequency ω.

The FWM induced cross talk is given by

$$P_{ijk}(L) = (\eta/9) D^2 \gamma^2 P_i P_j P_k e^{-\alpha L} \{[1 - \exp^{-\alpha L}]^2/\alpha^2\}$$

where P_i, P_j, and P_k are the input powers of the three input signals f_1, f_2, and f_3, L is the optical traveled length (i.e., fiber length), α is the attenuation coefficient, η is the FWM efficiency, D is a degenerative factor (equal to 3 for degenerative FWM or 6 for nondegenerative FWM), and γ is a nonlinear coefficient (for the fiber medium) given by

$$\gamma = (2\pi n_2)/(\lambda A_{eff})$$

where λ is the wavelength in free space, and A_{eff} and n_2 are the effective area and the nonlinear refractive index of the fiber, respectively.

5.6.3 Cross Talk

In closely spaced DWDM signals, adjacent channels influence each other, as in four-wave mixing, and optical power is transferred from one to the other through photon–matter–photon interactions. This power transfer is termed cross talk and is manifested as added noise on the signal. Since noise is cumulative, assuming that the cross talk noise-to-signal ratio (CSR) of adjacent channels in a link is δ_k, then for N links, the total CSR, δ_{tot}, is

$$\delta_{tot} = \Sigma \delta_k$$

for $k = 1$ to N. The cross talk noise degrades the quality of the received signal, which is manifested by eye closure, the penalty of which, $P_{cross\ talk}$, is estimated as

$$P_{cross\ talk} = -5 \log_{10}(1 - 2\sqrt{\delta_{tot}})$$

5.6.4 Polarization-Mode Dispersion

In polarization-mode dispersion (PMD), the two orthogonally polarized rays propagate at different group velocities, causing dispersion (measured in ps/km). The group delay (in ps) is expressed by:

$$\tau(\lambda) = \tau_0 + (S_0/2)\{\lambda - \lambda_0\}^2$$

In general, the differential group delay (DGD) for the length of fiber L is measured by the average difference of time arrival, $\Delta\tau$, between the two orthogonally polarized modes:

$$\Delta\tau = (\Delta n_g L)/c$$

where c is the speed of light and Δn_g is the refractive index variation corresponding to the group velocity of the orthogonal polarization states.

5.6.5 Stokes Noise and Chromatic Jitter

The fiber-core is not perfect. First, there are geometrical fluctuations so that its cross-section varies over its length, between circular and elliptical, in a nonuniform manner. Second, the dielectric constant varies nonuniformly over its length. Third, the optical channel is not purely monochromatic. Therefore, as an optical channel propagates in the fiber core, PMD is the unavoidable result of fiber-core birefringence and channel nonmonochromaticity. As a consequence, as two orthogonal (principal) states propagate in the fiber core, they experience a variability in polarization (due to nonuniform fluctuation of imperfections) known as *Stokes noise,* and variability due to optical-channel wavelength content, known as *chromatic jitter.* In this case, the square of the relative DGD noise $[d(\Delta\tau)/\Delta\tau]$ is the sum of squares of each noise component (Stokes-related and jitter-related):

$$[d(\Delta\tau)/\Delta\tau]^2 = [d(\Delta S)/\Delta S]^2 + [d(\Delta\omega)/\Delta\omega]^2$$

where $\Delta\omega$ is the frequency (or wavelength), d is the differential operator, and ΔS is the polarization state variability at the output of the fiber; the inverse of the latter is also known as the "bandwidth efficiency factor," α:

$$\alpha = 1/d(\Delta S)$$

The bandwidth efficiency factor is characteristic of the measuring method (it may have a value from under 1 to more than 200) and it is related to signal-to-noise ratio (SNR) of the optical signal in the following sense:

$$SNR \leq \alpha \Delta \tau \Delta \omega$$

That is, given the value of α, and measuring $\Delta \tau$ and $\Delta \bar{\omega}$, the potential maximum SNR is calculated, from which the potential amount of noise is determined.

5.6.6 Spectral Broadening

The refractive index depends on the electrical field. As a monochromatic pulse travels in a transparent medium, its amplitude variation causes *phase change* and *spectral broadening*. The *phase change* is given by

$$\Delta \Phi = [2\pi(\Delta n)L]/\lambda$$

where L is the fiber length, and Δn is the refractive index variation about the wavelength λ:

$$\Delta n = n(\lambda, E) - n_1(\lambda)$$

Phase variations are equivalent to frequency modulation or "chirping," which, as already described, is considered a form of photon shot noise.

The *spectral broadening* is given by

$$\delta \omega = -d(\Delta \Phi)/dt$$

For a *Gaussian*-shaped pulse, spectral broadening is

$$\delta \omega \doteq 0.86 \Delta \omega \Delta \Phi_m$$

where $\Delta \omega$ is the spectral width and $\Delta \Phi_m$ is the maximum phase shift in radians.

Spectral broadening appears as if one half of the pulse is frequency downshifted (known as *red shift*) and the other half is frequency upshifted (known as *blue shift*). Such shifts are also expected in pulses that consist of a narrow range of wavelengths that are centered on the zero-dispersion wavelength. Below the zero-dispersion point, wavelength dispersion is negative and above it positive. Significant spectral broadening is observed when $\Delta \Phi_m$ is equal to or greater than 2.

5.6.7 Self-Phase Modulation

The dynamic characteristics of a propagating optical pulse in fiber, due to the Kerr effect of the medium, result in modulation of its own phase, known as *self-phase modulation*. This nonlinear phenomenon causes spectral broadening.

If the wavelength of the pulse is below the zero-dispersion point (known as *normal dispersion regime*), then spectral broadening causes temporal broadening of the pulse as it propagates. If the wavelength is above the zero-dispersion wavelength of the fiber (the *anomalous dispersion regime*) then chromatic dispersion compensates self-phase modulation and reduces temporal broadening.

5.6.8 Self-Modulation or Modulation Instability

When a single pulse of almost monochromatic light has a wavelength well above the zero-dispersion wavelength of the fiber (the anomalous dispersion regime) another phenomenon occurs that further degrades the width of the pulse. That is, two side lobes are symmetrically generated at either side of the pulse, deforming the original pulse shape of the signal.

5.6.9 Attenuation

The optical attenuation coefficient of fiber, α, describes the optical power attenuation or loss per kilometer of fiber. Power loss degrades the quality of signal and decreases the signal-to-noise ratio as more noise is added on the attenuated signal. If the input and optical power at the entry and exit point of a 1 km long fiber is P_{in} and P_{out} (measured in mW), respectively, the optical attenuation over a fiber span L, L_{span}, is

$$L_{span} = \alpha L = 10 \log_{10}[P_{out}/P_{in}] \text{ (dB)}$$

Power loss mechanisms are additive and, thus, the total power is a straightforward algebraic summation. As an example, the total loss (including possible gain, G) in a path is calculated by the summation

$$P_\varepsilon - \Sigma L_n - \Sigma L_s - \Sigma L_c - \Sigma L_{N-L} - \Sigma L_{N-N} + \Sigma G_n$$

Where P_ε represents the power launched in the fiber, L_n the power loss of each fiber segment, L_s the mean splice loss, L_c the mean loss of line connectors, ΣL_{N-L} the sum of power degradations due to nonlinear effects, ΣL_{N-N} the sum of noise sources such as ASE, and ΣG_n the sum of amplifier gain (if applicable).

5.7 TEMPORAL PARAMETERS

The quality of optical signals is influenced by parameter fluctuations that can be of a linear or nonlinear nature. In addition, certain parameters vary with time, fast or slowly, which cause:

- Jitter of the modulated signal
- Temporal fluctuation of the spectral density of signal

- Temporal fluctuation of the optical power of signal
- Temporal fluctuation of the polarization state of signal

Jitter causes degradation of the received signal in the time domain, narrows the sampling window at the receiver, and is manifested by lateral eye-diagram closure (jitter is examined in more detail in Chapter 6). For instance, laser chirp under large signal stimulation is expressed by

$$2\pi \Delta f(t) = \alpha/2[dP/2dt + 2\pi K_P P - 2\pi K_{SP}/P]$$

where α is the linewidth enhancement factor, P is the laser output power, and K_P and K_{SP} are the contributions of nonlinear suppression and of spontaneous emission to adiabatic chirp.

Temporal fluctuation of spectral density is caused by laser drift and optical amplifiers. We have also discussed that when the wavelength is below the zero-dispersion point it causes temporal broadening of the pulse. In addition, it is well established that temperature, pressure, filter stability, and other conditions cause drift of the operating frequency. Variations of spectral density and operating frequency shift impact the dispersion estimations and channel spacing. Consequently, they impact the standard deviation of spectral distribution and the power level of the received signal. The frequency deviation is then calculated from

$$\Delta f(t) = -(c/\lambda_{mean}^2)[\lambda(t) - \lambda_{mean}]$$

where λ_{mean} is the mean wavelength of the laser source.

Temporal fluctuation of optical power is caused by lasers and by other components, including power variations caused during wavelength reassignment and add–drop. Power fluctuation impacts the power budget of the link and path.

Some temporal fluctuations are also attributed to ambient temperature variations, which affect the signal propagation (group velocity) and polarization states.

5.8 LINEAR, NONLINEAR, AND OTHER EFFECTS

The linear or nonlinear behavior of matter triggers mechanisms that affect the optical signal through it. However, one common effect that both linear and nonlinear matter has on optical power is attenuation. Attenuation is the effect of scattering and/or absorption of photons caused either by impurities or by disturbances in the uniform arrangement of atoms in the matrix of matter. In optical communications, this is manifested as power attenuation over the length of fiber or as insertion loss of components.

In addition to fiber attenuation, the optical signal experiences more passive power loss due to connectors, splices, and other components. Moreover, power is lost due to fiber-to-fiber mismatches, such as:

5.8 LINEAR, NONLINEAR, AND OTHER EFFECTS

- Connecting different fiber types
- Angular misalignment
- Core-diameter mismatch
- Lateral offset
- Numerical-aperture mismatch
- Concentricity mismatch
- Core-ellipticity mismatch
- Excessive distance (gap) between fibers
- Reflections at the end face of fibers, also known as *Fressnel reflections,* expressed as part of reflection loss

In particular, reflections may cause deleterious effects to the light source that cause frequency shift, OSNR decrease, and BER increase.

The dielectric constant of matter and, thus, its refractive index has a transfer characteristic that consists of a linear and a nonlinear region. Thus, a series expansion of it yields a first order (linear part) and higher-order terms (nonlinear):

$$n = n_0 + a_1 E^1 + a_2 E^2 + \ldots$$

Matter that has zero or negligible high-order terms is termed linear and matter for which high-order terms are nonzero is termed nonlinear. However, the refractive index is a function of the optical frequency. This dependence becomes more pronounced as matter is in the presence of strong fields.

Isotropic matter behaves linearly when it is free of stress (thermal, mechanical, or optical), and if the signal that traverses it has relatively low optical power. The linear behavior of matter causes:

- Phase shift of the optical wave that propagates through it
- Group velocity dispersion (GVD) as a result of phase shift variability between frequencies
- Dispersion as a result of speed variability between frequencies
- Spontaneous noise (optical amplifiers)

Matter with nonsymmetrical crystalline structure or under the influence of internal stress or high optical power interacts nonuniformly and nonlinearly with the electromagnetic wave nature of light. This nonlinear behavior of matter causes:

- Four-wave mixing (FWM)
- Self-phase modulation (SPM)
- Cross-phase modulation (XPM)
- Stimulated Brillouin scattering (SBS)
- Stimulated Raman scattering (SRS)

- Chromatic dispersion slope (S)
- Polarization-mode dispersion (PMD)
- Polarization-dependent loss (PDL)

Table 5.1 lists several interactions and the effect they have on the optical signal.

Power budget loss involves calculating all signal power degradations due to linear and nonlinear effects of the fiber, components, and amplifiers that affect the shape of the optical signal, its polarization states, its power level, and its noise and jitter content. A model of this is illustrated in Figure 5.9. Therefore, the optical power budget directly or indirectly involves all physical phenomena that affect the quality of the optical signal, such as

- Optical loss due to connectors, splices, patch panels, and other components
- Transmitter power and receiver sensitivity
- Amplification type, gain, and number of amplifiers in the path
- Length of each link
- Number of links in the path
- Fiber characteristics
- Expected received quality of signal (including BER and OSNR)
- Channel density (or channel separation) and channel width
- Linearity and nonlinearity effects (dispersion, FWM, etc.)
- Temperature variation along the fiber
- Modulation and bit rate
- And many more

Table 5.1. Cause and effect

Cause	Effect
λ interacts with λ	Interference
λs interact with matter	Linear and nonlinear effects: absorption, scattering, birefringence, phase shift, reflection, refraction, diffraction, polarization, polarization shift, PDL, modulation, self-phase modulation, etc.
λ–matter–λ interaction	FWM, issues, SRS, SBS, OFA
Nonmonochromatic channel	Pulse broadening, finite number of channels within available band
Refractive index variation (n)	Affects propagation of light
Transparency variation	Affects amount of light through matter
Scattering	Optical power loss (attenuation)
Reflectivity	Affects polarization of reflected optical wave
	Affects phase of reflected optical wave
Ions in matter	Dipoles interacting selectively with λs, energy absorption or exchange, affect refractive index

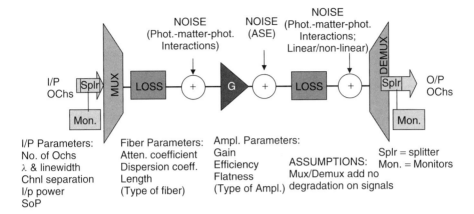

Figure 5.9. A Link model for the estimation of power loss, noise, and jitter on the signal.

5.9 PHOTODETECTOR RESPONSIVITY AND NOISE CONTRIBUTORS

The gain-current responsivity of a photodetector is given by

$$R = \eta eG/h\nu = \lambda \eta eG/hc$$

where η is the quantum efficiency, e is the charge of the electron (1.6×10^{-19} C), G is the gain of the detector, h is Planck's constant (6.63×10^{-34} joules-sec), $\nu = c/\lambda$ is the optical frequency, and c is the speed of light in free space ($\sim 3 \times 10^8$ m/s). Assuming a quantum efficiency $\eta = 0.5$, gain $G = 1$, and substituting the values of the constant quantities, the responsivity in terms of wavelength (measured in nm) is $R = \lambda \times 4.02 \times 10^{-4}$ mA/mW.

The received optical signal, prior to impinging the photodetector, consists of the original signal plus the sum of all noise and impairment contributors (Figure 5.10).

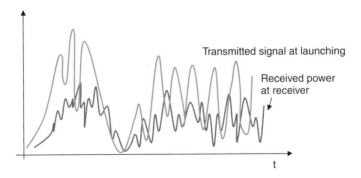

Figure 5.10. Effect of power loss and noise on an analog signal.

174 NOISE SOURCES AFFECTING THE OPTICAL SIGNAL

This is the signal that the photodetector converts in an electronic signal. However, the degradation of the signal is not over. The receiver itself adds its own share of noise, which consists of three contributors: dark current (N_{dark}), shot noise (N_{shot}) measured in A^2/Hz and (Johnson) thermal noise (NJ) measured in A^2/Hz (Figure 5.11):

$$N_{dark} = [2ei_d]B$$

$$N_{shot} = [2\eta e^2 P_{dif}/h\nu]B$$

$$N_{J-B} = [4kT/R_L]B$$

Where B is the receiver bandwidth, k is the Boltzman constant, T is the ambient temperature, i_d is the dark current of the photodetector, and R_L is the output load across whee the voltage is measured.

Now, the variance of the noise current per bit at the output of the receiver is:

$$\sigma^2_{N,photon} = 2(\eta P_R/h\nu)^2 \sigma^2_T$$

where P_R is the incident photonic power per bit and σ_T is the variance due to all contributing photonic and electronic noise variances, $\sigma^2_T = \Sigma \sigma^2_i$.

The contributing photonic and electronic noise variances, σ^2_i, receiver, dark current, optical amplifier ASE noise, filter, and relative intensity noise (RIN), are:

$$\sigma^2_{RCVR} = 4kTB/R_L$$

$$\sigma^2_{dark} = 2ei_d B$$

$$\sigma^2_{ampl} = 4R^2 P_R \rho_{ASE} B$$

$$\sigma^2_{filter} = 4gR^2 \rho^2_{ASE} B^2$$

$$\sigma^2_{RIN} = i_d^2 n_{RIN} B$$

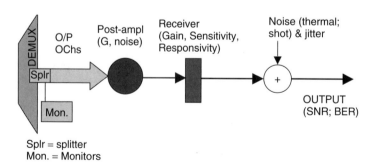

Splr = splitter
Mon. = Monitors

Figure 5.11. The receiver adds its own share of noise and jitter to an already corrupted optical signal.

where ρ_{ASE} is the spectral density of optical amplifier ASE, g is the ratio $\Delta\nu/B$, where $\Delta\nu$ is the optical filter bandwidth and B is the receiver bandwidth, and nRIN is the relative intensity noise (RIN) factor.

When RIN is assumed to be white and Gaussian with a two-sided spectral density, the RIN spectral power density is expressed (in Hz) as

$$S_{RIN} = 0.5 \times 10^{(RIN/10)} P_{Trans}^2$$

where *RIN* is the relative intensity noise parameter of the laser (measured in dB/Hz), and P_{Trans} is the average transmitted power.

Based on the above, the SNR is the ratio of the mean square value of the current generated by the receiver to the total noise variance in the photocurrent. In this case, the target SNR based on the minimum power and minimum number of photons required to detect a one bit (of a bit period τ_c) may be approximated by

$$SNR = (\eta P_R/h\nu)^2 / [(e^2 \eta P_R/h\nu + ei_d + 2kT/R_L)B + (e\eta P_R/h\nu)^2 \sigma_{N,ph}^2 + (e\eta P_R/h\nu)^2 \tau_c B]$$

Finally, if the receiver has a transimpedance amplifier (TIA), then the power spectral density of the TIA input noise is

$$psd_{TIA} = (I_n/R)^2 / 2n_{TIA}$$

where I_n is the input referred noise current of the TIA, the factor 2 appears because of the two-sided spectral density, and n_{TIA} is the noise bandwidth of the TIA defined as

$$n_{TIA} = (1/2\pi) \int |F_0(\omega)|^2 d\omega$$

where the integral is from 0 to infinity.

REFERENCES

1. M. Abramovitch and I. Stegun, editors, *Handbook of Mathematical Functions*, National Bureau of Standards, United States Department of Commerce, 1972.
2. S.M. Sze, *Physics of Semiconductor Devices*, Wiley, 1981.
3. C. Kittel, *Elementary Statistical Physics*, Robert E. Krieger, 1988.
4. M. Schwartz, *Information, Transmission, Modulation, and Noise*, McGraw-Hill, 1970.
5. A. Papoulis, *Probability, Random Variables, and Stochastic Processes*, 3rd ed., Mc-Graw-Hill, 1991.
6. A. Papoulis, *Systems and Transforms with Applications in Optics*, McGraw-Hill, 1968.
7. A. Yariv, *Optical Electronics*, 3rd ed., Chapter 11, pp. 345–382, CBS College Publishing, 1991.
8. S.O. Kasap, *Optoelectronics and Photonics*, Prentice-Hall, 2001.

9. R.J. Baker, *CMOS: Mixed Signal Circuit Design,* Wiley/IEEE Press, 2002.
10. S.V. Kartalopoulos, "*DWDM: Networks, Devices and Technology,*" Wiley/IEEE Press, 2003.
11. S.V. Kartalopoulos, *Fault Detactability in DWDM,* IEEE Press, 2001.
12. A. van der Ziel, *Noise: Sources, Characterization, Measurements,* Prentice-Hall, 1954.
13. S.V. Kartalopoulos, "What is DWDM?," *SPIE Optical Engineering Reports,* no. 203, pp. 4 and 12, November 2000.
14. E. Desurvire, *Erbium Doped Fiber Amplifiers,* Wiley, 1994.
15. F.N. Robinson, *Noise and Fluctuations in Electronic Devices and Circuits,* Oxford University Press, 1974.
16. M.J. Buckingham, *Noise in Electronic Devices and Systems,* Wiley-Halsted, 1983.
17. P.K. Pepeljugoski and K.Y. Lau, "Interfoerometric Noise Reduction in Fiber-Optic Links by Superposition of High Frequency Modulation," *Journal Lightwave Technology, 10,* 7, 957–963, June 1992.
18. I.T. Monroy, E. Tangdiongga, and H. de Waardt, "On the Distribution and Performance Implications of Filtered Interfoerometric Crosstalk in Optical WDM Networks," *Journal Lightwave Technology, 17,* 6, 989–997, June 1999.
19. I.T. Monroy, E. Tangdiongga, and H. de Waardt, "Interfoerometric-Crosstalk Reduction by Phase Scrambling," *Journal Lightwave Technology, 18,* 5, pp. 637–646, May 2000.
20. J. Zhang, M.Yao, Q. Xu, H. Zhang, C. Peng, and Y. Gao, "Interferometric Noise in Optical Time Division Multiplexing Transmission System," *Journal Lightwave Technology, 20,* 8, pp. 1329–1334, 2002.
21. B. Giles and E. Desurvire, "Modeling Erbium-Doped Fiber Amplifiers," *Journal Lightwave Technology, 9,* 271–283, February 1991.
22. E. Desurvire, J.R. Simpson, and P.C. Becker, "High-Gain Erbium-Doped Traveling-Wave Fiber Amplifiers," *Optics Letters, 12,* 11, 888–890,. November 1987.
23. R.J. Mears, L. Reekie, I.M. Jauncey, and D.N. Payne, "Low-Noise Erbium-Doped Fiber Amplifier Operating at 1.54 μm," *Electronics Letters, 23,* 19, 1026–1028, September 1987.
24. H. Meyr, M. Moeneclaey, and S.A. Fechtel, *Digital Communications Receivers,* Wiley, 1999.
25. F. Tong, "Multiwavelength Receivers for WDM Systems," *IEEE Communications Mag., 36,* 12, 42–49, November 1998.
26. J.B. Johnson, "Thermal Agitation of Electricity in Conductors," *Physical Review, 32,* 97–109, 1928.
27. H. Nyquist, "Thermal Agitation of Electric Charge in Conductors,"*Physical. Review, 32,* 110–113, 1928.
28. J. B. Johnson, "Electronic Noise, the First Two Decades," *IEEE Spectrum,* February 1971, pp. 42–46.
29. A. Papoulis, *Systems and Transforms with Applications in Optics,* McGraw-Hill, 1968.
30. G.P. Agrawal, *Nonlinear Fibwwer Optics,* Academic Press, 1995.
31. E. Desurvire, *Erbium Doped Fiber Amplifiers,* Wiley, 1994.
32. C.E. Shannon, "A Mathematical Theory of Communication," *Bell System Technical Journal, 27,* 379–423, 623–656, 1948.

33. S.O. Rice, "Mathematical analysis of random noise," *Bell System Technical Journal, 23,* 1–51, 1944.
34. N. Shibata, K. Nosu, K. Iwashita, and Y. Azuma, "Transmission Limitations Due to Fiber Nonlinearities in Optical FDM Systems," *IEEE Journal on Selected Areas in Communications, 8,* 6, 1068–1077, August 1990.
35. N.A. Olsson, and J. Hegarty, "Noise Properties of a Raman Amplifier," *IEEE Journal of Lightwave Technology, LT-4,* 4, 396–399, April 1986.
36. K. Vilhelmsson, "Simultaneous Forward and Backward Raman Scattering in Low-Attenuation Single-Mode Fibers," *Journal Lightwave Technology, LT-4,* 4, 400–404, 1986.
37. M. Yan, J. Chen, W. Jiang, J. Li, J. Chen, and X. Li, "Pump Depletion Induced Noise and Crosstalk in Distributed Optical Fiber Raman Amplifiers,"*IEEE Photonics Technology Letters, 13,* 7, 651–653, 2001.
38. J. Skinner and C.H.R. Lane, "A Low-Crosstalk Microoptic Liquid Crystal Switch," *IEEE Journal on Selected Areas in Communication, 6,* 7, 1178–1185, August 1988.
39. P.A. Humblet and W.M. Hamdy, "Crosstalk Analysis and Filter Optimization of Single- and Double-Cavity Fabry-Perot Filters," *IEEE Journal on Selected Areas in Communication, 8,* 6, 1095–1107, August 1990.
40. A.M. Hill and D.B. Payne, "Linear Crosstalk in Wavelength Division Multiplexed Optical Fiber Transmission Systems," *Journal Lightwave Technology, LT-3,* 643–651, 1985.
41. A.R. Chraplyvy, "Optical Power Limits in Multi-Channel Wavelength-Division-Multiplexed System due to Stimulated Raman Scattering," *Electronics Letters, 20,* 2, 58–59, January 1984.
42. B.E.A. Saleh and M.C. Teich, "Multiplied-Poisson Noise in Pulse, Particle, and Photon Detection," *Proceedings of IEEE, 70,* 220–245, 1982.
43. U. Fano, "Ionization Yield of Radiations. II. The Fluctuation of the Number of Ions," *Physical Review, 72,* 26, 1947.
44. R. Matloob, R. Loudon, M. Artoni, S. Bennett, and J. Jeffers, "Electromagnetic Field Quantization in Amplifying Dielectrics," *Physical Review A, 55,* 1623–1633, 1997.
45. J.R. Jeffers, N. Imoto, and R. Loudon, "Quantum Optics of Traveling-Wave Attenuators and Amplifiers," *Physical Review A, 47,* 3346–3349, 1993.
46. M.B. Gray, A.J. Stevenson, H.A. Bachor, and D.E. McClelland, "Harmonic Demodulation of Non-Stationary Shot Noise," *Optics Letters, 18,* 759–761, 1993.
47. T.M. Niebauer, R. Schilling, K. Danzmann, A. Ruediger, and W. Winkler, "Nonstationary Shot Noise and Its Effect on the Sensitivity of Interferometers," *Physical Review A, 43,* 5022–5029, 1991.
48. C.R.S. Fludger, V. Handerek, and R.J. Mears, "Fundamental Noise Limits in Broadband Raman Amplifiers," *Technical Digest OFC 2001,* paper MA5-1, March 2001.
49. P.C.D. Hobbs, "Shot Noise Limited Optical Measurements at Baseband with Noisy Lasers," in Laser Noise, R. Roy, editor, *Proc. Soc. Photo-Opt. Instrum. Eng., 1376,* 216–221, 1991.
50. A.M. Bacon, H.Z. Zhao, L.J. Wang, and J.E. Thomas, "Microwatt Shot-Noise Measurement," *Applied Optics, 34,* 24, 1995.
51. C.W.J. Beenakker and M. Patra, "Photon Shot Noise," *Modern Physics Letters B, 13,* 337–347, 1999.

52. H.P. Yuen and V.W.S. Chan, "Noise in Homodyne and Heterodyne Detection," *Optics Letters, 8,* 177–179, 1983.
53. H.A. Bachor and P.T.H. Fisk, "Quantum Noise—a Limit in Photodectors," *Applied Physics B. 49,* 291–300, 1989.
54. C.W.J. Beenakker and C. Schoenenberger, "Quantum Shot Noise," *Physics Today,* 37–44, May 2003.
55. A.L. McWhorter, "1/f Noise and Related Surface Effects in Germanium," in R.H. Kingston (Ed.): *Semiconductor Surface Physics,* University of Pennsylvania Press, 1956.
56. A. van der Zeil, "Flicker Noise in Electronic Devices," *Advances Electronics & Electron Physics, 49,* 225–297, 1979.
57. M.B. Weissman, "1/f Noise and Other Slow, Nonexponential Kinetics in Condensed Matter," *Review Modern Physics, 60,* 537–571, 1988.
58. S.B. Lowen and M.C. Teich, "Generalized 1/f Shot Noise," *Electronics Letters, 25,* 16, 1072–1074, 1989.
59. C. Schutte and P. Rademeyer, "Subthreshold 1/f Noise Measurements in MOS Transistors Aimed at Optimizing Focal Plane Array Signal Processing," *Analog Integrated Circuits and Signal Processing, 2,* 3, 171–177, 1992.
60. V. Radeka, "Low-Noise Techniques in Detectors," *Annals Review Nuclear Particle Sciences, 38,* 217–277, 1988.
61. P.S. Andre, L.L. Pinto, A.N. Pinto, and T. Almeida, "Performance Degradations due to Crosstalk in Multiwavelength Optical Networks Using Optical Add Drop Multiplexers Based on Fibre Bragg Gratings," *Revista Do Detua,* Portugal, 3, 2, 85–90, September 2000.
62. A.R. Chraplyvy, "High-Capacity Lightwave Transmission Experiments," *Bell Labs Technical Journal, 4,* 1, 230–245, 1999.
63. P.P.Mitra and J.B. Stark, "Nonlinear Limits to the Information Capacity of Optical Fibre Communications, *Nature, 411,* 1027–1030, 2001.
64. E. Narimanov and P.P. Mitra, "The Channel Capacity of a Fiber Optics Communication System," *Proceeding of Optical Fiber Communications Conference, 2002,* paper ThQ1, pp. 504–505, 2002.
65. W. Shieh, "Principal States of Polarization for an Optical Pulse," *IEEE Photonics Technology Letters, 11,* 5, pp. 677–679, 1999.
66. E. Lichtman, "Limitations Imposed by Polarization-Dependent Gain and Loss on All-Optical Ultralong Communications Systems," *Journal Lightwave Technology, 13,* 906–913, May 1995.
67. J.P. Gordon, and H. Kogelnik, "PMD Fundamentals: Polarization Mode Dispersion in Optical Fibers," *Proceedings of the National Academy of Sciences, 97,* 9, 4541–4550, April 25, 2000.
68. C.D. Poole, R.W. Tkach, A.R. Chraplyvy, and D.A. Fishman, "Fading in Lightwave Systems due to Polarization-Mode Dispersion," *IEEE Photonics Technology Letters, 3,* 68–70, January 1991.
69. C.D. Poole and R.E. Wagner, "Phenomenological Approach to Polarization Dispersion in Long Single-Mode Fibers," *Electronics Letters, 22,* 19, 1029–1030, 1986.
70. P. Lu, L. Chen, and X. Bao, "Polarization Mode Dispersion and Polarization Dependent Loss for a Pulse in Single-Mode Fibers," *Journal Lightwave Technology, 19,* 6, 856–860, June 2001.

71. P. Lu, L. Chen, and X. Bao, "Principal States Of Polarization for an Optical Pulse in the Presence of Polarization Mode Dispersion and Polarization Dependent Loss," in *Proc. 2000 International Conference on Applications of Photonic Technology* (ICAPT'2000), Quebec, Canada, June 12–16, 2000.
72. Y. Sun, A.K. Srivastava, J. Zhou, and J.W. Sulhoff, "Optical Fiber Amplifiers for WDM Optical Networks," *Bell Labs Technology J., 4,* 1, 187–206, 1999.
73. G. Eisenstein, "Semiconductor Optical Amplifiers," *IEEE Circuits and Devices Magazine, 5,* 4, 25–30, July 1989.
74. N. Gisin, "Statistics of Polarization Dependent Losses," *Optics Communications, 114,* 399–405, 1995.
75. G. Waarts, and R.P. Braun, "Crosstalk due to Stimulated Brillouin Scattering in Monomode Fibers," *Electronics Letters, 24,* 78–80, 1988.
76. P.T. Thomas, et al., "Normal Acoustic Modes and Brillouin Scattering in Single-Mode Optical Fibers," *Physical Review, B19,* 4986–4998, 1979.
77. K. Inoue, "Phase-Mismatching Characteristics of FWM in Fiber Lines with Multi-Stage Optical Amplifiers," *Optics Letters, 17,* 801–803, 1992.
78. K.O Hill, D. C. Johnson, B. S. Kawasaki, and R.I. MacDonald, CW Three-Wave Mixing in Single-Mode Optical Fibers," *Journal Applied Physics, 49,* 5098–5106, 1978.
79. J.P. Pocholle, J. Raffy, M. Papuchon, and E. Desurvire, "Raman and Four Photon Mixing Amplification in Single Mode Fibers," *Optical Engineering, 24,* 4, 600–608, 1985.
80. M. Yan, J. Chen, W. Jiang, J. Li, J. Chen, and X. Li, "Pump Depletion Induced Noise and Crosstalk in Distributed Optical Fiber Raman Amplifiers,"*IEEE Photonics Technology Letters, 13,* 7, 651–653, 2001.
81. J.R. Thompson and R. Roy, "Multiple four-wave mixing process in an optical fiber," *Optics Letters, 16,* 8, 557–559, April 1991.
82. K. Inoue, "Four-Wave Mixing in an Optical Fiber in the Zero-Dispersion Wavelength Region," *Journal Lightwave Tech., 10,* 11, 1553–61, November 1992.
83. K. Inoue, "Phase-Mismatching Characteristics of FWM in Fiber Lines with Multi-Stage Optical Amplifiers," *Optics Letters, 17,* 801–803, 1992.
84. K.O Hill, D. C. Johnson, B. S. Kawasaki, and R.I. MacDonald, CW Three-Wave Mixing in Single-Mode Optical Fibers," *Journal Applied Physics, 49,* 5098–5106, 1978.
85. C.R.S. Fludger, V. Handerek, and R.J. Mears, "Pump to Signal RIN Transfer in Raman Fiber Amplifiers," *IEEE Journal of Lightwave Technology, 19,* 8, 1140–1148, August 2001.
86. C.R.S. Fludger, V. Handerek, and R.J. Mears, "Fundamental Noise Limits in Broadband Raman Amplifiers," *Proceedings of Optical Fiber Communications Conference, 2001,* paper MA5-1, March 2001.
87. S. Song, C.T. Allen, K.R. Demarest, and R. Hui, "Intensity-Dependant Phase-Matching Effects on Four-Wave Mixing in Optical Fibers," *Journal of Lightwave Technology, 17,* 11, 2285–2290, November 1999.
88. J. Hansryd, H. Sunnerud, P.A. Andrekson, and M. Karlsson, "Impact of PMD on Four-Wave-Mixing-Induced Crosstalk in WDM Systems," *IEEE Photonics Technology Letters, 12,* 9, 1261–1263, September, 2000.
89. H. Kogelnik, "Theory of Dielectric Waveguides," in *Integrated Optics,* 2nd ed., Chapter 2, T. Tamir (Ed.), Springer-Verlag, 1979.

90. H. Kogelnik, "High-Capacity Optical Communications," *IEEE Selected Topics Quantum Elecelectronics, 6,* 6, 1279, 2000.
91. G. Bendelli, C. Cavazzoni, R. Giraldi, and R. Lano, "Optical Performance monitoring techniques," *Proceedings of European Conference on Optical Communication 2000, 4,* 113–116, 2000.
92. ANSI/IEEE 812-1984, *Definition of Terms Relating to Fiber Optics,* 1984.
93. IEC Publication 793-2, Part 2, *Optical Fibres—Part 2: Product Specifications,* 1992.
94. ITU-T Recommendation G.650, *Definition and Test Methods for the Relevant Parameters of Single-Mode Fibres,* 1997.
95. ITU-T Recommendation G.652, *Characteristics of a Single-Mode Optical Fiber Cable,* 1997.
96. ITU-T Recommendation G.653, *Characteristics of a Dispersion-Shifted Single-Mode Optical Fibre Cable,* 1997.
97. ITU-T Recommendation G.654, *Characteristics of a Cut-Off Shifted Single-Mode Optical Fibre Cable,* 1997.
98. ITU-T Recommendation G.655, *Characteristics of a Non-Zero-Dispersion Shifted Single-Mode Optical Fibre Cable,* 1996.
99. ITU-T Recommendation G.671, *Transmission Characteristics of Passive Optical Components,* 1996.

CHAPTER 6

Timing, Jitter, and Wander

6.1 THE PRIMARY REFERENCE SOURCE

Timing, like synchronization, is a very important function in all communications networks. A timing or clock signal is a periodic waveform (of a sinusoidal or almost square form) that is generated by the oscillator in the timing unit of a system or node. The timing signal provides the heartbeat based on which all functions in a node and the network operate in a timely manner, known as primary reference source (PRS). Thus, the precision of the timing signal is of extreme importance and it is specified by standards such as ITU-T Recommendation G.810.

The highest timing accuracy is known as "stratum 1." The most accurate clock is based on the element Cesium, which is calibrated by the National Bureau of Standards and provides the primary reference timing source. This clock provides a frequency reference of consistently high accuracy. This reference is broadcast over a GPS satellite link or over a CDMA wireless link and is received by an antenna that feeds a timing source. The timing source has a quartz-based or rubidium-based local oscillator that locks onto the received frequency and supplies a dual clock, a primary timing reference and a secondary one (for backup) to communications systems in the building at stratum 1 accuracy (as long as the local oscillator is locked onto the received frequency). The output clocks may be 8 kHz, 64 kHz, 1.544 MHz (DS1), 2048 kHz (E1), or composite (8 and 64 kHz). This is known as the building information timing supply (BITS) and it is described in ITU, ANSI, and Telcordia GR-378-CORE and GR-2830-CORE specifications for central offices as well as for controlled environment vaults (CEV).

Clock accuracy is expressed in parts per million or in stratum number, where stratum 1 is the highest accuracy (Table 6.1). In communications networks, even the minimum timing accuracy is 10^{-11}, or 2.5 slips per year; if a regular clock would slip 2.5 ticks per year, one would make a one-minute correction once in a lifetime.

ITU-T distinguishes synchronization clocks as type I to type VI, based on type and bit rate of network (more details may be found in ITU-T Recommendation G.812). In general, network providers may use in their synchronization plan the particular clock type that is best suited to their network topology and operational practices, and meets network performance objectives.

182 TIMING, JITTER, AND WANDER

Table 6.1. Clock accuracy by stratum level

Stratum	Minimum accuracy	Slip rate	Notes
1	10^{-11}	2.523/yr	Primary reference source (PRS)
2	1.6^{-8}	11.06/day	e.g., 4ESS/5ESS
3	4.6^{-6}	132.48/hr	e.g., 5ESS/DCS
4	3.2^{-5}	15.36/min	DCB/COT/DPBX/access*

*COT = Central Office Terminal; DCB = Digital Channel Bank; DPBX = Digital PBX.

- A type I clock is used at all levels of the synchronization of the digital hierarchy in 2048 kbit/s-based networks.
- Type II and III clocks are used in the synchronization of the digital hierarchy that includes the rates 1544 kbit/s, 6312 kbit/s, and 44,736 kbit/s.
- A type III clock is used in end offices that support the 1544 kbit/s digital hierarchy.
- A type IV clock is used in networks that support the 1544 kbit/s hierarchy.
- A type V clock is used in transit nodes of networks based on both 1544 kbit/s and 2048 kbit/s hierarchies, according to G.812, 1988 version.
- A type VI clock is used in existing local nodes of networks based on the 2048 kbit/s hierarchy, according to G.812, 1988 version.

6.2 THE PHASE-LOCK LOOP

The clock circuitry in a network element consists of a local oscillator, a phase-lock loop (PLL), which is locked on a reference clock. In general, although the accuracy of the PLL is the same as that of the reference clock when it is locked in step with it, it is not the same when it is free running. Thus, when it is locked with the reference clock it remains locked with it. Moreover, when framing synchronization is achieved, then the framer function, driven by the local PLL, flywheels on the incoming frames. However, occasionally, either the reference clock has intermittent lapses or the PLL drifts. Thus, there are several question of interest. In all, in communications networks there are three PLL parameters of interest: holdover accuracy, free-running accuracy, and pull-in/hold-in, which have been the subject of standards that define them and recommend parameter limits according to network topology, bit rates, and operational practices. Such parameters are:

Holdover stability is the amount of frequency offset that an oscillator has after it has lost its synchronization reference. Holdover stability has a limited duration, which is several hours. The holdover accuracy is lower than the PRS accuracy.

Free-running accuracy is the maximum fractional frequency offset of the oscillator when there is no reference clock and it is free running at its own accuracy, which is specified by the oscillator manufacturer. When a clock is not locked to a reference, the random noise components are negligible compared to deterministic effects like initial frequency offset.

Pull-in/Hold-in is the oscillator's ability to achieve or maintain synchronization with a reference. Its range is at least twice its free-running accuracy.

Based on these definitions and clock types (here we have greatly simplified the usage complexity among clocks by type; the interested reader should consult the ITU-T specifications):

- A type II clock has a more stringent holdover stability specification than a type I clock.
- Type I and type II clocks have a more stringent holdover stability specification than a type III clock.
- Thus, a type I clock has a more stringent holdover stability specification than a type III clock. However, type I clocks can also be deployed in 1544 kbit/s-based networks if their pull-in range, noise generation, and noise tolerance comply with type II and type III clocks for SDH compatibility.
- Similarly, type II clocks with type III clock holdover stability may be used in 2048 kbit/s-based networks if their noise generation, noise tolerance, and transient behavior comply with type I clocks when used in SDH.
- Finally, type V clocks are suitable for 2048 kbit/s-based SDH if their noise generation and short-term stability compy with type I clocks.

Network synchronization refers to method of information flow and start of frame identification. For example, in synchronous networks the start of the (information) frame is repetitive with a prefixed repetition rate; for example, SONET/SDH, DS1, and DS3. In all these, frames arrive seamlessly contiguous at the rate of 8,000 frames/sec. Conversely, in asynchronous networks such as IP, frames may not be seamlessly contiguous and that the start of (information) frame may not be repetitive. As such, synchronous and asynchronous networks both require timing (a clock) of certain accuracy. Nevertheless, synchronous networks are more sensitive to clock accuracy than asynchronous networks.

In asynchronous data networks, similar arguments hold. Data transmission is subject to all the timing impairments of synchronous transmission. The local clock must also meet an accuracy level, although its accuracy may be specified lower than that of the synchronous network. In current data networks, framing is not flywheeled. Each packet has a "start of frame" pattern that is recognized as the packet is received. However, once the "start of frame" has been identified, bit sampling at the physical (PHY) layer of the receiver is very similar to that of synchronous networks. However, as the network evolves to provide voice and real-time services, synchronization of the next-generation IP network also becomes important.

Thus, timing and synchronization issues, such as frequency variation during packet transmission and jitter, need to be considered, as frequency variation may cause buffer overflow or underflow of bits and, combined with jitter, result in missed bits or bits in error. As an example of buffer overflow/underflow, consider a binary signal at R bits/s traveling in a medium L meters long at a speed v. Then, the number of bits in transit at any time is LR/v bits. Thus, if $L = 100$ km, $R = 10$ Gbit/s, and the speed in the medium is $v = 2 \times 10^8$ m/s, then there are $(100 \times 10^3) \times (10 \times 10^9)/(2 \times 10^8)$ (m × bits/sec/m/sec) or 5,000,000 bits in transit.

Now, a temperature decrease of 1°F causes a 0.01% increase in propagation speed. This will cause 500 fewer bits in transit and, thus, possible buffer underflow. When the temperature increases, it may cause a similar overflow.

Thus, to meet transmission requirements frequency variation must be compensated for and jitter removed. One method of frequency variation compensation is addressed with a buffer commonly known as an elastic store. The elastic store smoothes out small variations of the incoming bit rate with respect to the local reference clock. In current all-optical networks, the elastic buffering cannot be cost-efficiently implemented with all optical components and, thus, the elastic store is placed at the far end of the path, at the receiver, where frequency compensation takes place.

In SONET/SDH, frequency variations of the incoming signal with respect to the reference clock are addressed with resynchronization at the output port and with pointer justification. In asynchronous data, this is addressed with long buffers.

6.3 BIT, FRAME, AND PAYLOAD SYNCHRONIZATION

In synchronous networks, such as SONET/SDH, at the input port of a node bit synchronization, frame synchronization, and payload synchronization are employed. Bit synchronization is the function that assures that each received bit has been decoded at the correct symbol value (1 or 0). Frame synchronization is a function that assures that the start of the frame has been located correctly. Payload synchronization is the function that identifies where the start of client data payload is in the frame.

Bit synchronization requires clock accuracy and precise sampling circuitry (in voltage threshold level and sampling instance). The sensitivity of the receiver, gain, decision threshold, and sampling instance are selected to decode the received symbols at an acceptable performance level.

Frame synchronization requires clock accuracy and a robust pattern detector and state machine to locate the start of the frame. In synchronous networks, frame synchronization is accomplished by detecting a periodic framing pattern in the header. Once the start of the frame is detected, the frame synchronizer flywheels on the repetitive frame. In asynchronous data networks, frame synchronization may be accomplished either as in synchronous networks or by first buffering and then identifying the frame header.

Payload synchronization requires a mechanism that locates the start of the client signal in the frame payload. In SONET/SDH, this mechanism is a pointer that identifies the offset between the start of the frame and the start of the payload in the frame. In DSn signals, it is accomplished in a deterministic manner by locating specified time slots of successive frames; each time slot contains a byte of the payload or digitized signal.

Bit synchronization is almost the same in all networks, synchronous or asynchronous. Frame synchronization, however, varies with communications protocol. In DS1, frame synchronization is accomplished with elastic buffers and with the framing "F" bit, which is distributed over 12 or 24 consecutive frames. In a 12-frame scheme, the pattern is 000111 (one bit every other frame), and in a 24-frame scheme (known as extended superframe), the pattern is 001011 (one "F" bit every four frames). In STS-1 SONET/SDH frames, the first two bytes of the frame are dedicated for frame synchronization. Once frame synchronization has been established, then payload synchronization is accomplished with the pointer bytes (bytes H_1 and H_2) and the action field (byte H3), both in the frame overhead.

In local area (data) networks the header precedes the information field, and in some protocols a preamble preceeds the header (such as fiber data distributed interface—FDDI), and, thus, the receiver identifies the start very quickly. In ATM, cell synchronization is accomplished using the head error control (HEC) byte. When data protocols (IP, ATM, and so on) are mapped onto SONET, then the SONET frame synchrinization comes first and then the IP or ATM start of frame in the payload (SONET overhead bytes indicate type of payload and pointers indicate the start of frame in the payload).

In conclusion, although the synchronization mechanism and state machine are not the same in all cases, the synchronization function and frame detection are extremely important in both synchronous and asynchronous networks.

6.3.1 SONET Synchronization—An Example

As a synchronization example, we examine a SONET add–drop multiplexing (ADM) node. Thus, a network element (NE) receives a reference clock from a stratum 1 BITS or from an OC-N signal, from the line or drop side (Figure 6.1). The timing requirements for this model are described in GR-436-CORE and ANSI T1.101 documents.

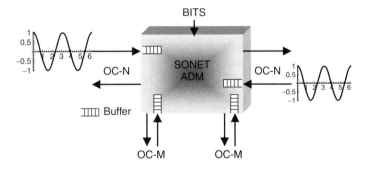

Figure 6.1. Synchronization model of a SONET ADM node.

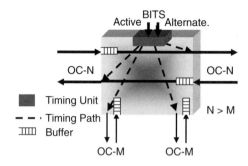

Figure 6.2. A BITS source provides timing to all output ports of the node.

At each signal input port, the signal is resynchronized to the reference frequency at all output ports. In SONET/SDH networks, the synchronization issue has been worked out with flexibility and it has been used as the synchronization paradigm for other modern networks. Thus, when a SONET NE is synchronized to an external protected BITS timing source (that is, there is an active and an alternate BITS timing input), then all outgoing signals of the ADM node are resynchronized to the BITS frequency accuracy (Figure 6.2).

If BITS is not available, then the node extracts timing from a designated DS-n in a OC-N from the line (OC-N) or from the add–drop side (OC-M) (Figure 6.3). This is known as line timing. In such cases, the extracted timing resynchronizes all output ports of the node.

In addition to these synchronization schemes, SONET specifies line timing synchronization from two (DS1) inputs, as well as loop timing, which is a special case of line timing. Loop timing has different holdover criteria and it applies to NEs that have only one OC-N interface (Figure 6.4).

As another special case, SONET/SDH defines through timing, which is particularly suitable for regenerators (Figure 6.5). This synchronization scheme requires two independent timing units that decouple the data signal in each direction; decou-

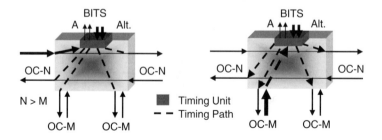

Figure 6.3. Extracted timing from a line OC-N or DS-n input signal (left) or from the drop side (right) becomes the timing source for this node.

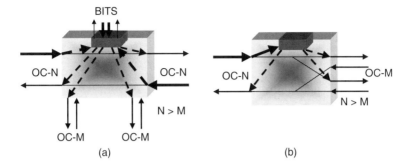

Figure 6.4. Line timing extracted from two DS1s (A), and loop timing (B) (buffering is not shown).

pling minimizes delay and eliminates additional buffering. However, this synchronization scheme is not recommended for ADM nodes.

Clock accuracy is measured at and depends on the timing hierarchical level; its accuracy must meet specified specifications. For example, the specified minimum free-running accuracy of a SONET minimum clock (SMC) is ±20 parts per million (ppm). If a SONET ADM is line timed, externally timed, or through timed and its reference signal fails it must enter the holdover state first. A digital cross-connect system (DCS) with external timing must have a minimum accuracy of ±4.6 ppm. Entry into holdover and restoration from holdover must be error free. GR-1244-CORE provides SONET oscillator timing specifications and ITU-T and other standards describe timing specifications for the SDH hierarchy (see references in this chapter).

In addition to clock accuracy, there is also clock stability. Clock stability is the subject of various standards. For example the SONET standard defines short- and long-term stability of OC-N signals, two cases of which are shown in Figure 6.6.

A key issue in networks is timing protection. If the node that provides timing for the other nodes in the network fails, or the link from it fails, then which node supplies timing? This is answered by having designated a priori an alternate node or link to assume the timing responsibility. Switchover, however, must be fast so that

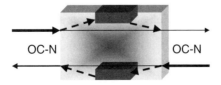

Figure 6.5. Through timing in regenerators requires two independent timing units, one per direction (buffering is not shown).

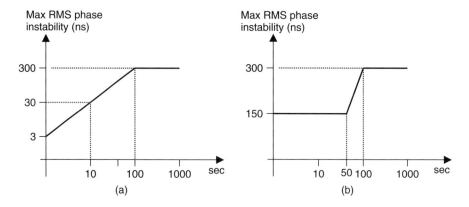

Figure 6.6. SONET defines short-term stability of the OC-N output (A), and short-term stability of the timing reference signal furnished by the network provider (B).

the oscillators do not free run. Clearly, when a failure occurs, then either the switchover action is autonomous or the result of a protocol.

6.4 SYNCHRONIZATION IMPAIRMENTS

Component degradations or failures and line parametric variations and impairments cause the receiver to loose its ability to synchronize. Timing depends on electronic phase lock loops that are frequency stabilized with crystals cut precisely to resonate at a particular frequency. The frequency accuracy of a crystal oscillator depends on material properties, geometry of the cut crystal, and operating temperature. However, even the most accurate phase-lock loop has a finite probability of error, which is measured in parts per million (ppm) or number of period slips from an expected one million periods.

A timing reference source may be interrupted for a period. For example, a satellite timing source may cease operation for a few or more minutes. In such cases, synchronization with the reference timing is lost, and the local oscillator enters the holdover state, where it may remain for up to eight hours. Another example is when the fiber that carries the synchronizing tributary (such as a DS1) is cut. If the loss of timing persists, then the oscillator enters the free-running state. In each state, the oscillator must meet certain accuracy requirements specified by standards.

The most common classification of impairments associated with synchronization are:

- Loss of signal (LOS)
- Loss of clock (LOC)
- Loss of frame (LOF)
- Loss of synchronization (LOS)

Loss of Signal (LOS) is caused by a long string of 0s or by a line cut. In this case, monitoring circuitry detects the LOS condition and the local oscillator goes into the holdover state, a state with relaxed frequency accuracy. For example, in SONET,

- If there is no light for 100 ms, then a LOS is declared.
- No LOS is declared if loss of light lasts for less than 2.3 ms.
- If there is no light for more than 2.5s, the NE sends an alarm message to the operating system (OS).

Loss of clock (LOC) is caused by a long string of 0s, a long string of 1s that exceeds the PLL holding time, or by a line cut. In such cases, circuitry detects the LOC condition and the local oscillator goes into a holdover state, a state with lessened frequency accuracy.

Loss of frame is results when there is excessive noise or excessive frequency variation between the local oscillator and the incoming signal as well as excessive jitter in the incoming signal. In such extreme occurrences, the framer detects LOF and enters a frame synchronization "hunt" state (in SONET/SDH). That is, it tries to find the start of frame again; in gigabit-Ethernet, the "hunt" state is not part of the synchronization state machine. In SONET, if there are excessive errors in the framing pattern due to excessive noise and jitter, and if a severely errored frame (SEF) persists for more than 3 msec and there are incorrect framing patterns for at least four consecutive frames, then the network element (NE) sends a message to the OS.

Loss of Synchronization results when there is excessive variation in frequency between the local oscillator and the received signal. Then the synchronizer goes into a LOS state and tries to resynchronize.

6.5 NETWORK HIERARCHY

The modern synchronous communications network consists of nodes interconnected with fiber. However, the requirements and functionality of a local access node that aggregates traffic to an OC-3 or OC-48 level cannot be the same as that of a massive cross-connect that passes the traffic of multiple OC-48 and OC-192s. As a consequence, a five-layer switching node hierarchy was established to differentiate functionality, switching requirements, aggregate traffic capacity, OA&M complexity, and synchronization. There are three principal types of interoffice trunks: *local* (interoffice, tandem, and intertandem), *toll,* and *intertoll.* The five layers are (Figure 6.7)

1. "End offices" (class 5),
2. "Toll centers" (class 4),
3. "Primary centers" (class 3),
4. "Sectional centers" (class 2),
5. "Regional centers" (class 1).

190   TIMING, JITTER, AND WANDER

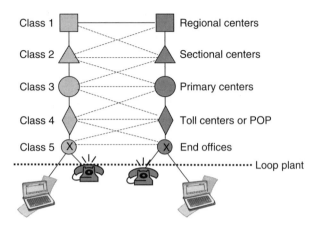

Figure 6.7. Network hierarchy.

In this hierarchy, synchronization accuracy requirements increase in importance from the end office to the higher-order office, or from the lower aggregate bandwidth (lower layer) to the higher aggregate bandwidth (higher layer) that demands a BITS with stratum 1 accuracy, whereas some others may have timing at the accuracy of a DS-n, received signal (Figure 6.8).

6.6 THEORETICAL FOUNDATION OF TIMING ERROR

In general, a timing signal $s(t)$ is mathematically described by the sinusoidal function (see ITU-T Recommendation G.810 for most of the definitions in this section)

$$s(t) = A \sin \Phi(t)$$

Where A is a constant amplitude coefficient and $\Phi(t)$ is the total instantaneous phase. The total phase $\Phi_{id}(t)$ of an ideal timing signal is expressed as

$$\Phi_{id}(t) = 2\pi \nu_{nom} t$$

Where ν_{nom} is the *nominal frequency*. However, because of parameter fluctuations, the phase of the signal generated by an actual oscillator is not perfectly periodic. That is, the intervals between successive equal phase instants are slightly different than the ideal expected phase. Based on this, the phase $\Phi(t)$ for actual timing signals is expressed as

$$\Phi(t) = \Phi_0 + 2\pi \nu_{nom}(1 + y_0)t + \pi D \nu_{nom} t^2 + \varphi(t)$$

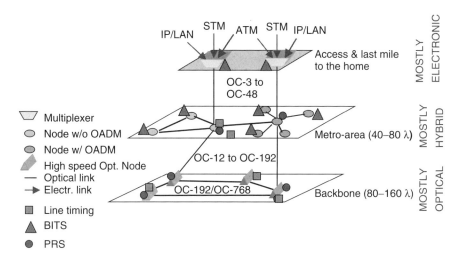

Figure 6.8. Depending on the layer, the clock accuracy and clock derivation may be different and may have different accuracies.

where Φ_0 is the initial phase offset, y_0 is the relative frequency offset variation from ν_{nom}, D is the linear frequency drift due to oscillator aging, and $\varphi(t)$ is a random phase variation.

The interval between two successive equal phase instants is also known as the clock "period" T and is inversely proportional to the clock frequency. This results in period and frequency fluctuations and timing errors of the actual timing signal with respect to ideal timing. Thus, the time function $T(t)$ of a realistic clock is the measure of ideal time t by the clock, and defined in terms of total phase and nominal frequency ν_{nom} as

$$T(t) = \Phi(t)/2\pi\nu_{nom}$$

The time deviation of the phase or period of an actual timing signal from the period of a reference timing source is called the time interval error (TIE) function and it is expressed as

$$TIE(t; \tau) = [T(t+\tau) - T(t)] - [T_{ref}(t+\tau) - T_{ref}(t)]$$

where τ is the observation interval.

TIE is measured in absolute values of units of time (such as seconds or nanoseconds). Depending on the variability of the actual clock, TIE may vary over some time linearly, sinusoidally, or randomly during an observation time ($\tau = n\tau_0$). The testing procedures for TIE are described by standards (e.g., ITU-T G.813, ITU-T O.172, ANSI T1.101, ETS 300 462-3).

Thus, in communication systems and networks, the time error of a clock is the difference between the time (or frequency) of this clock, $T(t)$, and the time (or frequency) of a reference clock, $T_{\text{ref}}(t)$. Thus, a time error function (TEF) $x(t)$ is defined as

$$x(t) = T(t) - T_{\text{ref}}(t)$$

The time error function serves in the calculation of a clock's stability parameters such as clock noise, wander, buffer characterisation, frequency offset and drift, and so on. Notice that TEF is a function of time and that its value may vary linearly, sinusoidally, or randomly. As a consequence, the TEF function depends on the timing model. Such models are the Maximum Time Interval Error (MTIE), the Allan Deviation (ADEV), the Modified ADEV (MDEV), the Time Deviation (TDEV), and the Root Mean Square of Time Interval Error (TIErms). We briefly describe the MTIE, TDEV, and ADEV; a description of all functions is beyond our purpose (for details, see ITU-T G.810, Appendix II).

6.6.1 Maximum Time Interval Error

The maximum peak-to-peak variation within an observation interval is called the maximum time interval error (MTIE) (Figure 6.9). MTIE is measured in time units and it is estimated by

$$\text{MTIE}(n\tau_0) \cong \max_{\langle 1 \leq k \leq N-n \rangle} [\text{peak-to-peak}]$$
$$= \max_{1 \leq k \leq N-n} \left[\max_{k \leq i \leq k+n} x_i - \min_{k \leq i \leq k+n} x_i \right], n = 1, 2, \ldots, N-1$$

The peak-to-peak MTIE value that occurs within an observation interval T may be used to calculate a relative frequency offset $\Delta f/f$ (for that interval) that is approximated to the ratio $|\text{MTIE}|/T$. Thus, if for an interval of 3s, MTIE is 21 µs, then the frequency offset is estimated as $21 \times 10^{-6}/3 = 7 \times 10^{-6} = 7$ ppm

6.6.2 Time Deviation

The time deviation (TDEV) is a measure of units of time of the expected time variation of a signal as a function of integration time. TDEV provides information

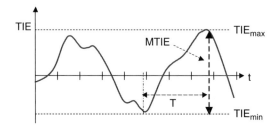

Figure 6.9. Definition of MTIE measured over an observation period.

about the spectral content of the phase noise and it is estimated based on the sequence of time error samples (ITU-T Rec. G.810, Appendix II, 8/1996):

$$\text{TDEV}(n\tau_0) = \sqrt{\frac{1}{6n^2}\left\langle\left[\sum_{i=1}^{n}(x_{i+2n} - 2x_{i+n} + x_i)\right]^2\right\rangle} = \frac{n\tau_0}{\sqrt{3}}\text{MDEV}(n\tau_0)$$

where x_i represents the time error samples, N is the total number of samples, τ_0 is the time error sampling interval, τ is the integration time, and n is the number of sampling intervals within the integration time τ. Thus, the integration time is $n\tau_0$. MDEV($n\tau_0$) is the modified Alan deviation, defined as

$$\text{MDEV}(n\tau_0) = \sqrt{\frac{1}{2(n\tau_0)^2}\left\langle\left[\frac{1}{n}\sum_{i=1}^{n}(x_{i+2n} - 2x_{i+n} + x_i)\right]^2\right\rangle}$$

Thus, an approximation of the TDEV is:

$$\text{TDEV}(n\tau) \cong \sqrt{\frac{1}{6n^2(N-3n+1)}\sum_{j=1}^{N-3n+1}\left[\sum_{i=j}^{n+j-1}(x_{i+2n} - 2x_{i+n} + x_i)\right]^2}$$

where $n = 1, 2, \ldots$ is the integer part of $N/3$.

To obtain the TDEV, one starts with obtaining TIE samples, the values of which are mapped on a plane with coordinates TIE and time (Figure 6.10A). From the TIE sample distribution, the standard deviations, $\sigma_x(s)$, and the time variance TVAR = σ_x^2, are obtained. Now, for all 1s intervals the root mean square (RMS) is obtained. This is repeated for all 2 s intervals, 3 s, and so on, and the RMSs are plotted against the intervals (1 s, 2 s, 3 s, and so on). The resultant is the TDEV curve (Figure 6.10B). The length of the interval is closely associated with the frequency of time interval error; thus, the slope of TDEV is related to noise. In fact, the shorter the interval, the higher the frequency of error is, and the longer the interval, the lower the interval (we will see that this is closely associated with jitter and the frequency of jitter). Therefore, TDEV helps in determining the short-term stability of the clock signal.

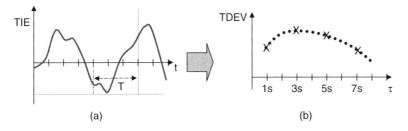

Figure 6.10. Obtaining the TDEV from the TIE.

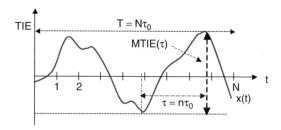

Figure 6.11. Definition of MTIE measured over an observation period T.

6.6.3 Maximum TIE

The maximum peak-to-peak value, x_{pp}, that occurs for all observations of length τ within T is known as maximum TIE [$MTIE(t)$]. The $MTIE(t)$ within an observation interval τ (Figure 6.11) may be used to calculate a relative frequency offset $\Delta f/f$ (for that interval), which is approximated by the ratio $|MTIE|/\tau$. Thus, if MTIE for an interval of 3 s is 21 μs, then the frequency offset is estimated to $21 \times 10^{-6}/3 = 7 \times 10^{-6} = 7$ ppm.

According to the definition of $MTIE(\tau)$, MTIE is independent of the sampling period τ_0, and based on offset estimates it can be used to characterize the buffer length. For example, if the buffer is designed to meet the worst-case MTIE, then the buffer will not overflow and bit/frame slips will be avoided as long as the MTIE value is not exceeded. Thus, MTIE helps in determining the long-term stability of the clock signal.

6.6.4 Allan Deviation

The Allan deviation $ADEV(\tau)$ is defined by

$$ADEV(\tau) = \sqrt{(1/2\tau^2)[x(t+2\tau) - 2x(t+\tau) + x(\tau)]^2}$$

where the argument within brackets denotes an ensemble average. Thus, $ADEV(n\tau_0)$ is estimated as

$$ADEV(n\tau_0) \cong \sqrt{\frac{1}{2n^2\tau_0^2(N-2n)} \sum_{i=1}^{N-2n}(x_{i+2n} - 2x_{i+n} + x_i)^2}$$

where $n = 1, 2, \ldots$ is the integer part $(N-1)/2$.

In terms of the power spectral density $S_\varphi(f)$ of the random phase deviation $\varphi(t)$ of a timing signal, the ADEV is

$$ADEV(\tau) = \sqrt{\frac{2}{(\pi\omega_{nom}\tau)^2} \int_0^{f_h} S_\varphi(f)\sin^4(\pi\tau f)df}$$

Table 6.2. Comparison of timing models

	Sampling period	Systematic effects (wander)	Buffer characterization	Clock noise
TDEV	Dependent	Sensitive	Not suited	Yes
ADEV	Independent	Sensitive	Not suited	Yes
MDEV	Dependent	Sensitive	Not suited	Yes
MTIE	Independent	Not as good as the other 3 models	Well suited	Not as good as the other 3 models

where ν_{nom} is the nominal frequency of the timing signal and f is the measured bandwidth.

ADEV(τ) is independent of the sampling period τ_0 and thus it is not influenced by a constant frequency offset (relative to the reference clock) of a timing signal. However, ADEV gives more information on the clock noise than MTIE, it is not suitable for buffer characterization, and it is sensitive to systematic effects, such as wander.

6.6.5 Comparison of Timing Models Defined in G.810

The behavior of four timing models is examined in terms of dependency on sampling period τ_0, sensitivity to systematic effects (such as wander), suitability to characterize buffer size, and if they provide information on clock noise. This comparison is summarized in Table 6.2.

6.7 TRAFFIC PROBABILITY AND SIGNAL QUALITY

In legacy synchronous communications networks, call arrival, call duration, and so on are probabilistic quantities. For example, call arrival approximates a *Poisson* distribution (Figure 6.12):

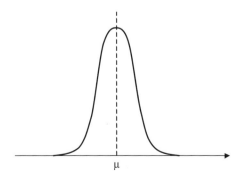

Figure 6.12. Poisson distribution.

$$P = e^{-A}\Sigma A^x/x!, \qquad x = n \text{ to infinity}$$

where A is the expected traffic density, n is the number of channels, and x is a variable describing the number of busy channels.

In asynchronous data communications, this probabilistic behavior repeats itself, but in this case we have packet arrival, packet length (and, thus, packet duration), and so on.

Thus, if there are n samples, each with amplitude x_n, these samples are statistically distributed or dispersed around a mean value μ. The dispersion of a distribution, known as the *standard deviation*, σ, is calculated from the weighted square of differences as

$$\sigma = \sqrt{\{1/(n-1)\}\Sigma(x_i - m)^2}$$

The square of the dispersion, σ^2, is known as the *variance* of the sample, and the *variance-to-mean* ratio is known as *coefficient of overdispersion*, α:

$$\alpha = \sigma^2/\mu$$

Three cases are of interest here:

1. When $\alpha > 1$, traffic is termed *rough*.
2. When $\alpha < 1$, traffic is termed *smooth*.
3. When $\alpha = 1$, traffic is termed *random*.

Although traffic per se is not an obvious cause of jitter and bit errors, the probabilistic and statistical behavior of it is, particularly in networks with optical add–drop multiplexers. For example, traffic is related to effective bandwidth, which in WDM optical communications is calculated from the bit rate (a fixed quantity per lightpath), packet rate (a statistical quantity), and optical channel (OCh) density (a fixed or dynamic variable). Thus, taking into account the variability in power loss per path and OCh density, one can deduct that the statistical behavior of traffic impacts signal noise level, signal jitter, signal cross talk, and errored bits in the signal.

6.8 JITTER AND WANDER

We have seen that the frequency of actual clocks does not remain precise over time, but it varies fast or slow, positively or negatively, causing timing deviations, the effects of which have been studied with various models. These inaccuracies may cause bit, frame, or payload slips and thus loss of customer data. As a consequence, nodes in communications networks require a high degree of synchronization accuracy between the clock of a node and the incoming signal. For instance, if a node is synchronized to stratum 1 and the arriving signal is drifting or it arrives from a system with a free-running clock, then it is clear that there will be a small frequency deviation between the two. As a result, the arriving data may be slightly faster or slower than the reference clock. If it is slower, the reference clock may slip from time to time. If it is faster, the node may periodically lose incoming bits as it is not able to cope with a faster incoming rate. In either case, a defense mechanism must exist so that incoming data is buffered, retimed, or its rate adjusted with the clock of the node. This adjustment is known as "frequency justification," as already described in the SONET/SDH example. In this case, when the incoming clock has excessive jitter, known as high-frequency jitter, then fast justifications are required. Fast justification becomes another source of jitter (at the outgoing data) known as "pointer jitter."

Assume two timing sources are at the same frequency. Assume also that one of the two is a reference timing source and the other is a clock derived from incoming data, and, thus, it may drift with respect to the reference source. However, depending how fast or slow the frequency deviation changes, it may cause different effects. Therefore, a distinction is made between "fast" and "slow" variation that results in two outcomes: jitter and wander. In telecommunications, jitter and wander are very important and have been the subject of long-term research as well as standards documents (such as, ITU-T, ANSI, and Telcordia).

Wander is defined as the very slow variation between a reference frequency and a derived frequency (from incoming data). Per ITU-T Recommendation G.810 ("Definitions and Terminology for Synchronization Networks," 8/96), wander is defined as "the long-term variations of the significant instants of a digital signal from their ideal position in time (where long-term implies that these variations are of frequency less than 10 Hz)." A similar definition also is given for SONET (GR-253-CORE), which also puts the frequency range for wander at below 10 Hz.

Per G.810, jitter is "the short-term variation of a signal's significant instants from their ideal position in time," where short-term implies that these variations have a frequency greater than or equal to 10 Hz. Thus, 10 Hz seems to be the demarcation frequency variation below which it is defined as wander, and above which it is defined as jitter; this is also the demarcation frequency variation in the North American hierarchy (DS1-DS3).

In SONET, jitter requirements that pertain to asynchronous DS-N interfaces to NE are category I, and requirements that pertain to OC-N, STS-N electrical and synchronous DS1 interfaces to NE are category II.

Jitter requirements apply to interfaces between carriers and users, and between two carriers. Compliance with jitter requirements assures that jitter limits at OC-N and STS-N interfaces are met. Jitter and wander drive the specifications (G.823, G.824, and G.825) for synchronization requirements in SDH.

In general, jitter or wander have a different effect on signal quality, on system design, and on the propagation of jitter throughout the network. Depending on the mechanism that causes jitter or wander, it may be random, linear (triangular), or sinusoidal (Figures 6.13 and 6.14). As a side note, the profile of jitter or wander depends on the accuracy of the reference timing source.

Because jitter is the result of fast frequency variations, a practical unit needs to be defined for a measure of jitter amplitude with respect to the period of the signal. However, jitter has a different time value for different rates and, therefore, such units should be independent of the clock frequency and bit rate signal coding and be

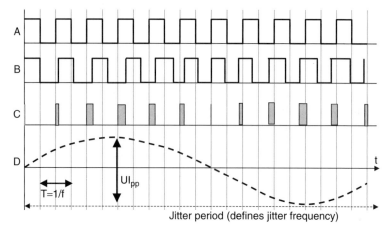

Figure 6.13. Explanation of sinusoidally periodic jitter. The actual "jittery" clock (B) is compared with a perfect clock (A). The amplitude evolution of sinusoidal jitter (C) is plotted in (D), where jitter period and jitter frequency are identified.

dimensionless. Such a unit is the Unit Interval (UI). Thus, a UI is a measure of the peak-to-peak deviation of a real clock from a reference clock over a predefined measuring interval, and it can be as long as the period of the reference clock. Thus, a UI may also be viewed as the normalized period of the reference clock. For example, in SDH (ITU-T G.783), the UI values for different STM-N/OC-Ns (Table 6.3) are provided (notice that these values are the period of the reference clock).

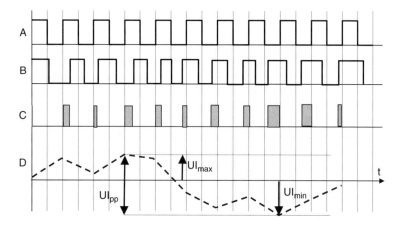

Figure 6.14. Explanation of random jitter. The actual "jittery" clock (B) is compared with a perfect clock (A). The random jitter evolution (C) is plotted in (D). The UI max–min and UI peak-to-peak are defined.

Table 6.3. Unit intervals for STM-N/OC-N

STM-1/OC-3	1 UI = 6.43 ns
STM-4/OC-12	1 UI = 1.61 ns
STM-16/OC-48	1 UI = 0.40 ns
STM-64/OC-192	1 UI = 0.10 ns
STM-256/OC-768	1 UI = 0.025 ns

In Figures 6.13 and 6.14, in addition to the unit interval (UI) parameter, the UI peak-to-peak and the root-mean-square (RMS) jitter are defined. Peak values are instantaneous values and over an observation interval there may be several peak-to-peak values. $UI_{pp,max}$ is the maximum of the UI_{pp}.

The RMS jitter value represents the average power of jitter within the interval of observation. RMS jitter is defined by the expression

$$J_{RMS} = \sqrt{(1/T)\int [f(t)]^2 dt}$$

where $f(t)$ is a description of the time signal of jitter and the integral is from 0 to T (the interval of observation). Notice that $f(t)$ may represent a random, deterministic, or periodic function (such as a sinusoidal).

6.8.1 Types of Jitter

In general, jitter is classified as random or deterministic. Random jitter (RJ) may be attributed to one or more sources such as optical-amplifier noise, single-sideband noise, photon–matter–photon interaction in the fiber medium, and random refractive index and core variability over the length. The generation of random jitter is approximated to a Gaussian probability distribution function expressed as

$$P[s(t)] = (1/\sqrt{2\pi\sigma^2})\exp[-1/\{2([s(t)-s_s]/\sigma)^2\}]$$

where $s(t)$ is the measured signal at time t, s_s is the mean value of the density function, and σ is the standard deviation.

Deterministic jitter (DJ) is attributed to several causes such as SONET/SDH framing and pointer justifications, duty cycle distortions, and initial frequency offset (when a clock from free-running tries to lock in a reference clock). This type of jitter, being deterministic and not random, cannot be described by distributions. DJ is further classified as

- Intersymbol interference (ISI) and data-dependent jitter (DDJ)
- Pulse-width-distortion jitter (PWDJ)
- Sinusoidal jitter (SJ)
- Uncorrelated bounded jitter (UBJ)

6.8.1.1 Intersymbol Interference (ISI).
ISI data-dependent jitter is the outcome of a number of optical impairments due to photon–matter–photon interaction and due to dependence of the dielectric constant and the refractive index on frequency. The latter causes dispersion so that symbols interact with their neighboring symbols. However, the contribution of ISI to jitter may be negligible compared to other jitter sources.

6.8.1.2 Data-Dependent Jitter (DDJ).
The pattern-dependent jitter (PDJ) is caused by repetitive patterns, such as the SONET/SDH A1 (Hex F6) and A2 (Hex 28) bytes in the SONET/SDH frame, and it can be explained as follows (Figure 6.15).

The A1 and A2 are fixed NRZ codes that are not scrambled. Therefore, these bytes repeat every 125 ms (8 kFrames/s) and, thus, the jitter generated by the bits in the A1 and A2 bytes should have an 8 kHz spectral band. All other bytes in the STN-N frame are scrambled and, thus, the jitter generated by them is random. Thus, the jitter component generated by A1 and A2, because of periodicity, has a more distinctive spectral band than the jitter component of other bytes in the frame. As such, in the case of an OC-N signal, jitter should peak at a high frequency, which is related with the OC-N rate (such as 2.5, 10, or 40 Gbit/s) with the peak having a spectral band of 8 kHz because of the frame periodicity. For example, in an OC-192 (10 Gbit/s) signal, PDJ peaks at 3.24 MHz within a spectral band of 8 kHz.

6.8.1.3 Pulse-Width Distortion Jitter (PWDJ).
If we consider that an OOK modulator does not generate a square or symmetric trapezoidal pulse, then the optical pulse is neither symmetric nor do the rise and fall times have the same slope. As

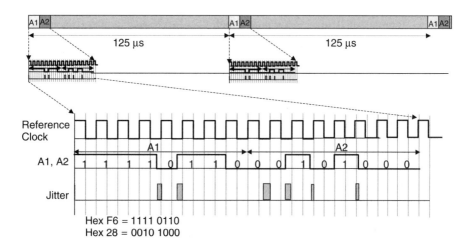

Figure 6.15. Explanation of pattern-dependent jitter of the A1 and A3 sychronization overhead bytes in an OC-3 signal.

such pulses propagate, the jitter induced at the edges of the pulses has a different distribution at each edge.

6.8.1.4 Sinusoidal Jitter (SJ).
SJ is caused by sinusoidal variations of timing sources that couple energy in electrically controlled devices that affect the optical signal. This jitter is also known as repetitive, and it may also be caused by switching power supplies that affect the transmitter or receiver.

6.8.1.5 Uncorrelated Bounded Jitter (UBJ).
UBJ may be caused by power supply variations, unpredictable cross talk, secondary transmission phenomena, and, in general, sources that are not described in the previous classifications. For example, bursts of jitter may be caused by cross talk with bursty data and uncorrelated FWM data interference. It may also be caused by intermittent failures, power or phase-lock-loop glitches, and timing slips.

In conclusion, understanding the mechanisms that generate jitter helps to better understand the impact it has on the quality of signal, and bit error rate BER.

6.8.2 Signal Affected by Jitter

Jitter affects the overall quality of signal at the receiver in three ways:

1. Stability of the rise and fall times of optical pulses
2. Stability of the rise and fall slopes of optical pulses
3. Stability of the width of the optical pulses

As a result of excessive jitter, the receiver creates erroneous bits, and/or the buffers at the receiving port may overflow, losing bits.

At the receiver, sampling is performed at a specific point in time, approximately near the 50% point of the bit period. Thus, if the sampling point is at the 0.5 point of the clock period, and the jitter amplitude is at or more than 50% of the unit interval (UI), then the sampled bits may be erroneous. Thus, jitter contributes to bit error rate (BER).

6.8.3 Sources of Jitter

Many sources contribute to the overall jitter of the signal. In DWDM signals, such sources are:

- Transmitter clock instability causes jitter or slow jitter (wander).
- The chirping effect of optical transmitters causes jitter.
- Optical receivers may add noise and jitter due to filter, transimpedance amplifier, photodetector noise, and receiver clock instabilities.
- In a SONET/SDH periodic signal, justifications and pointer adjustments due to small variations of the client data with respect to the SONET/SDH fram-

ing clock cause jitter, as do timing variations in tributaries (payload), such as DS-ns.
- Fiber nonlinear interactions with the optical signal (such as cross talk, XPM, FWM, PMD).
- Interactions between bits of the same signal (such as intersymbol interference)
- Amplifier random noise, such as ASE. ASE may affect the rise/fall time of optical pulses and thus cause jitter. Amplifier jitter is cumulative.
- Multiplexing and mapping may cause jitter.
- Duty cycle distortions may cause jitter.
- Stuffing and variable delay of asynchronous data jitter.
- Nonmonochromatic optical sources.
- Repeater jitter (in 3R optical–electrical–optical systems) is cumulative, and in long-haul systems where there are several repeaters, jitter may reach the maximum permissible level; this is known as "jitter peaking."

6.8.4 Examples of Self-Induced Optical Noise and Jitter: Stokes Noise and Chromatic Jitter FWM

When optical fiber is made, it is made to conform to parameter specifications recommended by standards. However, the values of fiber parameters, and particularly of the fiber core, exhibit parameter variations along the length of the fiber; this variation typically is random. Although parameter variations may be unnoticeable over few hundreds of meters of fiber, it is noticeable over many kilometers, and as light propagates, parameter variations affect its propagation. To envision this, consider a long road with small bumps randomly dispersed along it. When a vehicle travels at low speed for a very short distance, such bumps may not be noticed. However, when it travels at high speed and a long distance, then they will be very noticeable. Using this paradigm, as light travels along the fiber, even small variations of the fiber parameter will affect its propagation and the photon–matter interaction.

When a monochromatic signal propagates in the fiber, it is separated into states of polarization because of fiber birefringence, each state traveling at different speed and phase, thus causing polarization-mode dispersion (PMD). However, even PMD experiences random polarization variability as the two orthogonal states propagate in the fiber core because of nonuniform fluctuations of fiber-core imperfections. This is known as *Stokes noise*.

When a polychromatic signal from an actual optical source propagates in fiber, the dependence of the dielectric on wavelength causes a temporal pulse broadening or chromatic dispersion (CD). However, a random variation of the fiber dielectric along its axis causes a random variation of CD, manifested as noise and jitter, known as *chromatic jitter*.

The random variation of the fiber dielectric affects the propagation properties of light in many ways, generating different noise contributions; the sum of each noise contributor is manifested as differential group-delay (DGD) noise. The square of the relative DGD noise is the sum of squares of the Stokes noise and the chromatic-jitter-related noise:

$$[d(\Delta\tau)/\Delta\tau]2 = [d(\Delta S)/\Delta S]2 + [d(\Delta\overline{\omega})/\Delta\overline{\omega}]2$$

where d is the differential operator, $\Delta\overline{\omega}$ is the frequency variability, ΔS is the polarization-state variability at the output of the fiber, and $\Delta\overline{\omega}$ is the frequency variability, The inverse of the latter is known as the bandwidth-efficiency factor, α. The bandwidth efficiency factor depends on the measuring method and its value is estimated empirically; this value may be from under 1 to more than 200. The value of α provides an indication of the potential maximum SNR of the optical signal as

$$SNR \leqq \alpha\Delta\tau\Delta\overline{\omega}$$

In addition to Stokes and CD noise, there are other noise-generating mechanisms due to fiber-parameter variations, such as self-phase modulation (SPM) noise and modulation-instability (MI) noise. The square of these noise sources also add to the overall relative DGD noise in a general equation such as

$$N_{DGD}^2 = N_{STOKES}^2 + N_{CD}^2 + N_{SPM}^2 + N_{MI}^2$$

6.8.5 Examples of WDM Optical Noise and Jitter: Four-Wave Mixing

In addition to self-inflicted noise by a single optical channel, in DWDM there are additional noise and jitter contributions due to the photon–matter–photon interactions among channels. For example, as described in Chapter 5, two or three optical channels in close spectral proximity (such as 25 or 12.5 Ghz) interact to generate a fourth wavelength; this is known as four-wave mixing (FWM), λ_{FWM}. However, the λ_{FWM} product is modulated by three generating channels, the data of which are uncorrelated in the time domain. As a result, the λ_{FWM} carries random noise; we call this FWM data-induced noise. In addition, there is a secondary FWM noise contribution due to the dielectric variation along the fiber core. This variation modulates the intensity of the λ_{FWM} product randomly, adding to the FWM data-induced noise. The λ_{FWM} interferes with another data carrying channel, which is in the same spectral band, and it superimposes its noise power onto it.

6.8.6 Jitter Generation, Tolerance, and Transfer

Jitter generation, jitter tolerance, and jitter transfer are three jitter characterizatics of interest in communications systems design and performance..

Jitter generation includes sources of jitter, the mechanisms by which jitter is coupled with the signal, statistical properties of jitter, and jitter models for the probabilistic estimation of jitter.

Jitter tolerance provides a measure of the maximum amount of jitter that can be present without causing an error.

Jitter transfer is the amount of jitter that passes through a device, from input to output. When jitter exceeds the permissible limit, either synchronization could be lost or bit errors could be increased. Thus, a jitter transfer function (JTF) is defined to describe the amount of jitter (in dB) as a function of jitter frequency that passes from input to output through the system:

$$JTF(f) = 20 \log [J_{out}(f)/J_{in}(f)]$$

where J_{in} and J_{out} are jitter in and jitter out, respectively.

Knowing the JTF of a system, an assurance about the signal quality at the receiver is provided. However, the actual JTF only has meaning if a reference JTF is defined. Such reference is provided by standards that define a cutoff frequency (since the JTF acts like a low-pass filter) (Figure 6.16).

Because "fast variations" or "fast jitter" are fuzzy terms, communications standards define how fast is "fast" and what the limits of the expected jitter are in the form of templates. Such limits are provided in UI_{pp} and in a specific frequency range for a given STS/STM rate. For example, ITU-T G.783 specifies (G.783, Table 0-1) that the UI_{pp} jitter for STM-64 optical (10 Gbit/s) should be 0.3 UI_{pp} in the range 20 kHz to 80 MHz or 0.1 UI_{pp} in the 4 to 80 MHz range. Similarly, for a STM-256 optical (40 Gbit/s) it should be 0.3 UI_{pp} in the 80 kHz to 320 MHz range or 0.1 UI_{pp} in the 16 MHz to 320 MHz range.

6.8.7 Jitter Filtering Templates

In ultrafast optical transmission, it is reasonable to expect that the jitter requirements for 40 Gbit/s and for 155 Mbit/s are different, since jitter for these two cases impacts the optical signal differently. Therefore, a jitter tolerance for each case

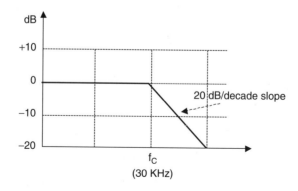

Figure 6.16. An example of a JTF template. The number in parentheses defines the cutoff for OC-3.

needs to be established, as well as a relationship between network interface jitter requirements and input jitter tolerance. That is, how much jitter an input port should tolerate and what is the maximum allowable jitter from the network (clearly, the network jitter should be less than jitter tolerated). Therefore, the SDH standards (such as ITU-T G.783, G.825, and O.172) define two high-pass jitter filters, a low-pass filter (based on a 60-second measuring interval), and their corresponding breakpoint frequencies (f_1, f_3, and f_4 respectively) for each STM-M/OC-N rate (Figure 6.17). Per ITU-T O.172, the nominal breakpoint frequencies are at −3 dB and the maximum attenuation is at least 60 dB. In addition, the measuring accuracy of jitter depends on the frequency response and the pseudorandom test pattern sequence.

6.8.8 Maximum Jitter Tolerance Templates

All input ports of communications nodes are required to tolerate (or accommodate) some degree of jitter, optical or electrical, without losing synchronization or increasing bit errors. However, the maximum tolerable jitter (MJT) is not the same for all tributary types and for all bit rates, and, therefore, there are standards that specify the maximum tolerable jitter in each case. These standards provide tables or a graphical template (or mask) that plots maximum permissible jitter boundaries in terms of UI versus jitter frequency (Figure 6.18). These templates are used in jitter conformance tests of inputs.

Communications systems must operate within the specified jitter tolerance so that they do not suffer error performance degradation. As a consequence, system jitter tests are performed (following the procedures specified by standards) to measure the actual jitter tolerance and plot the measured values on the MJT template. From this plot, one can deduct whether the jitter performance is acceptable or unacceptable and also determine the operating jitter margin (Figure 6.19). The jitter margin is also necessary for those cases in which the oscillator enters the hold over or the

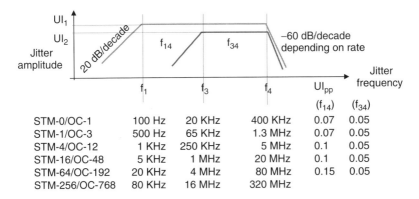

Figure 6.17. SDH/SONET breakpoint frequencies for measuring line jitter, and the maximum peak-to-peak jitter error (UI_{pp}) for structured signals in the bandwidths f_1–f_4 and f_3–f_4.

206 TIMING, JITTER, AND WANDER

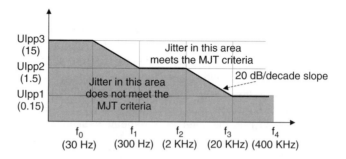

Figure 6.18. An example of a MTJ template. When the measured jitter crosses the boundary, then jitter has exceeded the maximum permissible jitter on the signal. In such cases, jitter measurements would have failed the jitter test. The values within parentheses are for the OC-3 case.

free-running state (for which the clock accuracy is allowed to be less than for the locked-in state).

6.8.9 Bit Error Ratio Penalty Criterion

Per ITU-T, "the bit error ratio (BER) penalty criterion for jitter tolerance measurements is defined as the amplitude of jitter, at a given jitter frequency, that duplicates the BER degradation caused by a specified Signal-to-Noise Ratio (SNR) reduction." To measure this penalty, ITU-T recommends that two tests be performed with a test apparatus that is able to control jitter, SNR, and measure BER. First, the signal noise or attenuation is increased (with jitter absent) until a convenient initial BER is obtained. Then, the noise or attenuation is decreased until the SNR at the decision circuit meets the expected value. Second, jitter (in a given frequency band) is added to the signal until the BER returns to its initial value. This step is repeated for the number of frequency bands needed to cover the applicable frequency range.

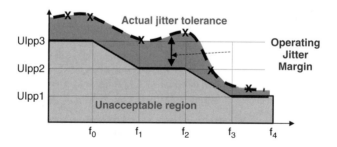

Figure 6.19. The actual jitter tolerance is plotted against the MJT to ensure that jitter requirements are not violated and to establish the operational jitter margin.

6.9 WANDER

In Section 6.8, wander was defined as the very slow variation between a reference frequency and a derived frequency (from incoming data). Per ITU-T Recommendation G.810 ("Definitions and Terminology for Synchronization Networks," 8/96), wander is defined as "the long-term variations of the significant instants of a digital signal from their ideal position in time (where long-term implies that these variations are of frequency less than 10 Hz)." A similar definition also is given for SONET (GR-253-CORE), which also puts the frequency range for wander at below 10 Hz.

Thus, the mechanisms for generating jitter, studied in section 6.8 are almost the same but they evolve very slowly in the time domain, as in "slow motion." Because of the slow evolution of wander (<10 Hz), wander testing is a time-consuming process that requires hours or days to complete. For example, wander of 0.1 Hz frequency corresponds to a testing interval of $1/0.1 = 10$ seconds. However, wander can be a small fraction of hertz. Thus, wander of 0.00001 Hz (slow, yet realistic) corresponds to $1/10^{-5}$ seconds or 1 day and 4 hours.

Because wander is a very slow variation, unlike jitter, the practical unit is nanoseconds and not UI.

6.9.1 Sources of Wander

In DWDM communications, the sources that contribute to wander are many. For example, slow gradual environmental variations, pointer variations, parametric variability of fiber, and component aging are a few.

6.9.2 Signal Affected by Wander

We described wander as a slow variation of jitter (< 10 Hz). Fast clocks in optical communications cope very easily with the slow variation of wander, which does not contribute to bit error or buffer overflow compared with jitter. When data from many sources are multiplexed, even though the magnitude of wander of each source may not cause problems, there may be many frequency justifications, since these sources do not track each other, that pointer jitter may eventually result. In addition, wander is cumulative, as it is transferred by amplifiers/regenerators in a cascade. Thus, excessive wander may contribute to jitter, which puts a limit on the number of amplifiers/regenerators on the path and, thus, on the length of the path. In fact, ITU Recommendations G.811, G.812, and G.813 provide limits for cumulative wander of transmission paths and for cumulative wander in synchronization distribution paths with cascaded clocks. Although wander is a slowly varying phenomenon, these requirements also set the maximum permissible wander at the output of synchronization nodes, the maximum expected wander at the input ports, wander specifications for primary and secondary (or master and slave) clocks under locked and free-running conditions, and wander transfer. To summarize, the wander requirements related to system design are:

- Maximum permissible wander at the output ports of nodes
- Maximum wander tolerance at the input ports of nodes
- Wander for primary reference clock or for the slave clock. This includes:
  Intrinsic output wander when the PLL is in the locked condition
  Intrinsic output wander when the PLL is in the free-running condition
  Wander transfer characteristics

However, wander limits are specified according to one of the timing models. For example, for PRS and for the MTIE model, G.811 defines wander MTIE limits as:

$(0.275 \times 10^{-3}\tau + 0.025 \; \mu s)$ for the observation interval $0.1 < \tau \leq 1000$ s
$(10^{-5}\tau + 0.29 \; \mu s)$ for the range $\tau > 1000$ s

For the slave clock, when in PLL lock condition, the MTIE wander limits for a type I node clock are (G.812 also specifies limits for type II, III, and IV clocks, not listed here):

24 ns for the observation interval $0.1 < \tau \leq 9$ s
$(8 \times \tau^{0.5}$ ns) for the observation interval $9 < \tau \leq 400$ s
160 ns for the observation interval $400 < \tau \leq 10{,}000$ s

Similarly, for the PRS for the TDEV model, the wander TDEV limits are:

3 ns for the observation interval $0.1 < \tau \leq 100$ s
$0.03 \; \tau$ ns for the observation interval $100 < \tau \leq 1000$ s
30 ns for the observation interval $1000 < \tau < 10{,}000$ s

For the slave clock, when in PLL lock condition, the TDEV wander limits for type I node clock are (G.812 also specifies limits for type II, III, and IV clocks, not listed here):

3 ns for the observation interval $0.1 < t \leq 25$ s
$0.12 \; \tau$ ns for the observation interval $25 < t \leq 100$ s
12 ns for the observation interval $100 < t \leq 10{,}000$ s

Similarly, the wander tolerance limits for the MTIE models (for a type I clock) are:

0.75 μs for the observation interval $0.1 < t \leq 7.5$ s
$0.1 \; \tau$ μs for the observation interval $7.5 < t \leq 20$ s
2 μs for the observation interval $20 < t \leq 400$ s
$0.005 \; \tau$ μs for the observation interval $400 < t \leq 1000$ s
5 μs for the observation interval $1000 < t \leq 10{,}000$ s

REFERENCES

1. D. Middleton, *An Introduction to Statistical Communication Theory,* IEEE Press, 1996.
2. P. R. Triscitta and E. Varma, *Jitter in Digital Transmission Systems.* Artech House, 1989.
3. P.R. Trischitta, and P. Sannuti, "The Accumulation of Pattern-Dependent Jitter for a Chain of Fibre Optic Regenerators," *IEEE Transactions on Communications, 36,* 6, 761–765, June 1988.
4. E.G. Shapiro, M.P. Fedoruk, and S.K. Turitsyn, "Numerical Estimate of BER in Optical Systems with Strong Patterning Effects," *Electronics Letters, 37,* 19, 1179–1181, 2001.
5. F. Matera and M. Settembre, "Role of Q-factor and of Time Jitter in the Performance Evaluation of Optically Amplified Transmission Systems," *IEEE J. of Selected Topics in Quantum Electronics, 6,* 2, 308–316, 2000.
6. A. Yariv, "Signal-to-Noise Considerations in Fiber Links with Periodic or Distributed Optical Amplification," *Optic Letters, 15,* 19, 1064–1066, October 1990.
7. K. Fussgaenger and S. Casselli, *Per Channel Bit Error Ratio, Q-Factor, EI, SNR and OSNR Relations for Single- and Multi-Channel Long-Haul Systems with Optical Amplifiers,* in Draft Recs. G.691, G.692, and G.onp, ITU-T SG 15 White Paper Contribution, COM 15-31, November 1997.
8. H. Okamura, *Asynchronous Sampling Technique for the Q-Factor Measurement in Transparent Optical Networks,* ITU-T SG 15 Delayed Contribution, NTT, D.393 (WP 4/15), Q 16/15, October 1998.
9. T. Alberty and V. Hespelt, "A New Pattern Jitter Free Frequency Error Detector," *IEEE Transactions on Commuications, 37,* 2, 159–163, February 1989.
10. R.G. Kusyk, W.A. Krzymien, and T.E. Moore, "Analysis of Techniques for the Reduction of Jitter Caused by SONET Pointer Adjustments," *IEEE Transactions on Commuications, 42,* 2036–2050, February/March/April 1994.
11. R.P. Singh, S-H. Lee, and C-K. Kim, "Jitter and Clock Recovery for Periodic Traffic in Brodband Packet Networks," *IEEE Transactions on Communications, 42,* 5, 2189 2196, May 1994.
12. J.Y. Hui, E. Karasan, J. Li, and J. Zhang, "Client-Server Synchronization and Buffering for Variable Rate Multimedia Retrievals," *IEEE JSAC, 14,* 226–237, January 1996.
13. K. Murakami, "Jitter in Synchronous Residual Time Stamp," *IEEE Transactions on Communications, 44,* 6, 742–748, June 1996.
14. O. Collins, "The Design of Low Jitter Hard Limiters," *IEEE Transactions on Communications, 44,* 5, 601–608, May 1996.
15. J. Walker and A. Cantoni, "Modeling of the Synchronization Process Jitter Spectrum with Input Jitter," *IEEE Transactions on Communications, 47,* 2, 316–324, February 1999.
16. M. Stefanovic, N. Zlatkovic, and D. Milosevic, "The Probability of Error in AM-PM Systems Due to Noncoherent Demodulation, Interference and Noise," *IEEE Transactions on Communications, 42,* 1445–1449, February/March/April 1994.
17. L.C. Schooley and G.R. Davis, "Jitter Effects on Bit Error Rate Reexamined," *IEEE Transactions on Communications, 29,* 6, 920–922, June 1981.

STANDARDS

1. ANSI T1.101, *Syncronous Interface Standards for Digital Networks*, 1999.
2. ANSI T1.102, *Digital Hierarchy–Electrical Interfaces*, 1993.
3. ANSI T1.105, *Synchronous Optical Network (SONET)—Basic Description Including Multiplex Structures, Rates and Formats*, 1995.
4. ANSI T.105.01, *Synchronous Optical Network (SONET)—Automatic Protection Switching*, 1995.
5. ANSI T1.105.03, *Telecommunications—Sychronous Optical Network (SONET)—Jitter at Network Interfaces*, 1994.
6. ANSI T1.105.03a, *Synchronous Optical Network (SONET)—Jitter and Network Interfaces—DS1 Supplement*, 1995.
7. ANSI T1.105.03b, *Synchronous Optical Network (SONET)—Jitter and Network Interfaces—DS3 Supplement*, 1995.
8. CCITT Recommendation G.703, *Physical/Electrical Characteristics of Hierarchical Digital Interfaces*, 1991.
9. ITU-T Recommendation G.707, *Network Node Interface for the Synchronous Digital Hierarchy (SDH)*, 1996.
10. ITU-T Recommendation G.735, *Characteristics of Primary PCM Multiplex Equipment Operating at 2048 Kbit/s and Offering Synchronous Digital Access at 384 Kbit/s and/or 64 KBit/s*, 1988.
11. ITU-T Recommendation G.742, *Second Order Digital Multiplex Equipment Operating at 8448 KBIit/s and Using Positive Justification*, 1988.
12. ITU-T Recommendation G.751, *Digital Multilex Equipment Operating at the Third Order bit rate of 34,368 KBit/s and the Fourth Order bit rate of 139,264 KBit/s and Using Positive Justification*, 1988.
13. ITU-T Recommendation G.781, *Synchronization Layer Functions*, 6/1991.
14. ITU-T Recommendation G.783, *Characteristics of Synchronous Digital Hierarchy (SDH) Equipment Functional Blocks*, 10/2000.
15. ITU-T Recommendation G.810, *Definitions and Terminology for Synchronization Networks*, 8/96.
16. CCITT Recommendation G.811, *Timing Requirements at the Outputs of Primary Reference Clocks Suitable for Plesiochronous Operation of International Digital Links*, 1988.
17. CCITT Recommendation G.812, *Timing Requirements at the Outputs of Slave Clocks Suitable for Plesiochronous Operation of International Digital Links*, 1988.
18. ITU-T Recommendation G.813, *Timing Characteristics for SDH Equipment Slave Clocks (SEC)*, 1996.
19. CCITT Recommendation G.822, *Controlled Slip Rate Objectives on an International Digital Connection*, 1988.
20. INCITS TR 25-1999, *Fiber Channel, Methodologies for Jitter Specification*, 1999.
21. ITU-T Recommendation G.823, *The Control of Jitter and Wander Within Digital Networks which are Based on the 2048 kbit/s Hierarchy*, 1993.
22. ITU-T Recommendation G.824, *The Control of Jitter and Wander Within Digital Networks which are Based on the 1544 kbit/s Hierarchy*, 1993.

23. ITU-T Recommendation G.825, *The Control of Jitter and Wander Within Digital Networks which are Based on the Synchronous Digital Hierarchy (SDH)*, 1993.
24. ITU-T Recommendation G.832, *Transport of SDH Elements on PDH Networks—Frame and Multiplexing Structures*, 10/1998.
25. ITU-T Recommendation G.955, *Digital Line Systems Based on the 1544 kbit/s and the 2048 kbit/s Hierarchy on Optical Fibre Cables*, 1993.
26. ITU-T Recommendation G.957, *Optical Interfaces for Equipments and Systems Relating to the Synchronous Digital Hierarchy*, 1995.
27. ITU-T Recommendation G.958, *Digital Line Systems Based on the Synchronous Digital Hierarchy for Use on Optical Fibre Cables*, 1994.
28. ITU-T Recommendation G.8251, *The Control of Jitter and Wander within the Optical Transport Network (OTN)*, 11/2001.
29. ITU-T Recommendation O.150, *General Requirements for Instrumentation for Performance Measurements on Digital Transmission Equipment*, 1996.
30. ITU-T Recommendation O.171, *Timing Jitter and Wander Measuring Equipment for Digital Systems which are Based on the Plesiochronous Digital Hierarchy (PDH)*, 4/1997.
31. ITU-T Recommendation O.172, *Jitter and Wander Measuring Equipment for Digital Systems which are Based on the Synchronous Digital Hierarchy (SDH)*, 3/2001.
32. Telcordia (formerly Bellcore) GR-378-CORE, *Generic Requirements for Timing Generators, Issue 2*, February 1999.
33. RFC 1305, *Network Time Protocol (version 3), Specification, Implementation and Analysis*.
34. Telcordia (formerly Bellcore) GR-253-CORE, *Synchronous Optical Network (SONET) Transport Systems: Common Generic Criteria*, Issue 2, December 1995.
35. Telcordia (formerly Bellcore) GR-1377, *SONET OC-192 Transport Systems Generic Criteria*, Issue 3, August 1996.
36. Telcordia (formerly Bellcore) TR-NWT-499, *Transport Systems Generic Requirements (TSGR): Common Requirements*, issue 5, December 1993.
37. Telcordia (formerly Bellcore), TR-NWT-917, *Regenerator*, October 1990.

CHAPTER 7

Probability Theory of Bit Error Rate

7.1 INTRODUCTION

As data is transmitted over a medium, attenuation, combined noise, and jitter sources all distort the shape of the transmitted bits, both in amplitude and time, to such a degree that a receiver misinterprets some bit values and detects them wrongly; that is, some logic "ones" are detected as logic "zeros" and some logic "zeros" as logic "ones." In communications, the number of error bits in the number of bits transmitted provides a performance metric of the channel, from the transmitter to (and including) the receiver. However, this metric needs clarification. For example, if two data rates are 1 Mbit/s and 10 Gbit/s, 10 errors in a second mean $10/1,000,000$ (or 10^{-5}) and $10/10,000,000,000$ (or 10^{-9}) errors, respectively. Alternatively, 10 errors in 1,000,000 bits transmitted means 10 errors per second for the 1 Mbit/s rate and 100,000 errors per second for the 10 Gbit/s rate.

Thus, depending on performance limits set for a specific application, the channel performance may or may not be acceptable. That is, the frequency (or rate) of erroneous bits is very critical. Although it is impossible to predict if a particular bit will be received correctly or not, it is possible to predict with good confidence the performance of a channel if the parameters of the link are known, as well as the statistical behavior (Gaussian, Poisson) of noise and jitter sources. Then, the frequency of occurrence of erroneous bits and the signal-to-noise ratio can be reliably estimated. What we have stated without having defined yet are the bit error ratio and the bit error rate. What they are and what the difference between the two is is examined in the next section. Thus, to model a transmission channel, a thorough knowledge is required of the link from transmitter to receiver, including the transmission medium and all components in between (Figure 7.1), as well as the sources of noise and jitter (including linear and nonlinear multiwavelength interactions) and the laser and photodetector characteristics.

In previous chapters, we discussed the light source and receiver, the medium, loss and gain, and noise and jitter. In this chapter, our attention is focused on how these degrading sources impact the value of a binary bit and change it from "one" to "zero," and "zero" to "one." We estimate the probability of errors and we provide an estimation methodology that can be implemented with an integrated solid-state circuit, thus allowing for a continuous estimation at each port, and making cumbersome measuring instruments necessary only for out-of-service precision testing.

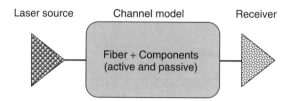

Figure 7.1. A channel is modeled from source to receiver with all impairments between, including fiber loss, nonlinearity, active/passive components, and noise and jitter sources.

The bit error rate and the optical signal quality are simulated in several exercises using the CD-ROM that accompanies this book (see Appendix B for a description of these exercises).

7.2 BIT ERROR RATIO AND BIT ERROR RATE

In the literature, one encounters two somewhat confusing terms. The first is the bit error ratio (BERatio), defined as the number of received bits in error over a large number of bits transmitted. The second term is the bit error rate (BERate) defined as the ratio of errored bits to the total bits transmitted in a time interval. Now, the term "ratio" by itself is static and it does not imply time, whereas the term "rate" implies time. For example, a performance objective of 10^{-11} BERatio means 1 error in 100,000,000,000 bits received, which at 1 Gbit/s will take 100 seconds to be observed. Contrary to this, the same objective of 10^{-11} BERatio for 40 Gbit/s will take 2.5 sec to be observed. The point is that 10^{-11} by itself does not explicitly define the performance metric unless the bit rate is also stated, in which case BERatio and BERate become equivalent, so that 10^{-11} means 1 error in 100,000,000,000 bits transmitted at a bit rate of xGbit/s and, hence, BER. As we already discussed, another consequence of bit error rate is the different interval required to observe the performance metric for different bit rates, such as the two bit rates of 10 and 40 Gbits/s (e.g., 10 versus 2.5 seconds).

Another consequence is that we have assumed a continuous data stream, as is the case in synchronous communications. Clearly, in asynchronous data (packet) communications, there may be idle time (or idle packets) between packets that carry customer or control and maintenance data. Thus, the term *bit error ratio* may be more meaningful in this case to distinguish the asynchronous nature of data transmission. In such case, one looks into the packet and counts total bits and total erroneous bits in it to calculate the ratio. However, even in this case the bit rate on the line (or transmission medium) needs to be stated.* Thus, although BER is a good measure of channel performance, it is not indicative of the perfor-

---

*Notice that sometimes the data rate is not the same as line rate. For example, although the data rate may be 1 Gbit/s, due to 4B/5B or 8B/10B coding the line rate is 1.25 Gbit/s

mance during idle intervals, if there are bursts of errors, or if errors are normally distributed over time.

Therefore, for a known bit rate, one calculates the number of errored bits in a short unit of time to provide a good measure of the bit error rate (BER) and its variation over time. In fact, by assuming a sliding time window, the channel error rate profile and statistical error behavior are better understood, such as error distribution and bursts of errors. In fact, most measuring instruments work this way, making the bit error rate more meaningful.

A direct method to calculate BER is by using bit error detection and correction codes (EDC). However, although EDCs are routinely used because of their corrective features, they also have their own inaccuracies and limitations and they require long observation times. We will examine EDCs in a subsequent chapter.

In this chapter, we examine the probability and statistics of errored bits and we estimate the OSNR, Q-factor, and OBER. These parameters are also important not only for the the overall quality of signal but also for the characterization of the optical channel and its performance.

7.3 DEFINITIONS

The bit error rate is an important metric for the performance characterization of a transmission channel. The BER performance metric is applicable to all types of transmission media, channels, and modulation methods. Media include wired, atmospheric, ionized, almost-free space, and fiber optic. Channels may be single, multifrequency, multiwavelength, time division multiple access, random and so on. Modulation methods include electrical, wireless, optical amplitude modulation, frequency-shift keying, phase-shift keying, multilevel, and several others. As a consequence, equipment or circuitry required for BER measurements and channel characterization is classified based on complexity, form factor, accuracy, and cost in service or out of service.

Out-of-service BER is measured using pseudorandom bit sequences (PRBS) or bit patterns that are generated by specialized instruments. Such PRBS patterns have a maximum sequence length of $2n - 1$, where n is a large odd number (usually 21, although other lengths have been specified); randomness minimizes pattern interference with data scramblers. During the measurement time, service for the channel under test is interrupted.

In-service BER is measured using error detecting/correcting (EDC) codes that have been embedded in the actual data stream. Such EDCs are Cyclic Redundancy Check (CRC), Parity Check, and observation of Bit Interleaved Parity (BIP). EDCs detect one or more errors in a given sequence of bits. However, EDCs are capable of detecting and correcting a limited number of errors and errors above their limit are not detectable or correctable; in the latter category are bursts of errors that exceed the limits of the EDC. EDCs also require long periods of BER computation time.

The probabilistic approach of estimating BER is another method that can be used out of service and particularly in service. This method, which will be further analyzed in this chapter, provides a qualitative estimate of SNR, Q-factor, and BER. We will see that this method estimates the channel performance in a fraction of time as compared with the time required for EDC methods.

Errored bits in a given sequence of bits, known as a *block,* can be measured in several ways. Standards provide several definitions for these errors:

- *Errored block* (EB) is a block with at least one errored bit.
- *Block error ratio* (BER) is the ratio of blocks with at least one bit error to the total number of blocks transmitted in a given time interval. For small values, the block error ratio is comparable to bit error ratio, and for specific error models it is possible to calculate the bit error ratio from the block error ratio.
- *Errored second* (ES) is a one-second period with at least one EB.
- *Severely errored second* (SES) is a one-second period in which more than 30% of the blocks have errors.

Among the error performance parameters are:

- *Errored second ratio* (ESR) is the ratio of ES to total seconds of available time during a fixed measurement interval.
- *Severely errored second ratio* (SESR) is the ratio of SES to total seconds in a fixed measurement interval.

Standards also provide the end-to-end error performance objectives, the full description of which is beyond our purpose. The emphasis here is that BER is seriously considered in all types of communications.

7.4 OSNR AND SPECTRAL MATCHING

The amount of optical power of noise that is mixed with the optical power of signal is indicative of the channel transfer characteristic. A channel performance parame-

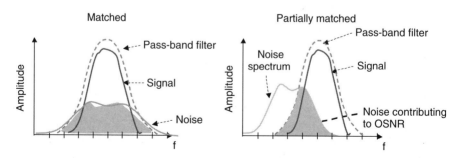

Figure 7.2. The optical signal-to-noise ratio depends on how well noise matches the spectral distribution of signal and filter. Compare the diagram on the left with that on the right.

ter that is used in such cases is the ratio of RMS signal power to RMS noise power. This is known as the *optical signal-to-noise ratio* (OSNR). It is important to notice that both optical signal and optical noise in this ratio need be distributed in the same spectral range, and that the optical filter at the receiver needs to match the spectral range of the signal and noise. Noise power is proportional to the filter bandwidth. Noise outside the spectral range of the signal and filter does not contribute to OSNR of the particular signal (Figure 7.2). When the signal is converted into an electrical signal, then the term "optical" is dropped and the more generic term signal-to-noise ratio (SNR) is considered.

7.5 CARRIER-TO-NOISE RATIO

Prior to discussing this topic, we have to go back in time to when digital transmission was under intense study and many powerful formulae were developed that set the foundation of modern digital transmission—optical, wired, or wireless. At that time, a major issue under study was the conversion of an analog signal to binary, so that when it is reconstructed back to analog it is a faithful replica of the original analog signal with minimal or unnoticed distortion. Shannon and Nyquist were among those who developed the maximum channel capacity and the sampling criterion of interest in this section.

Today, it is well known that in order to digitize an analog sample, V_s volts in n bit binary code, since there are 2^n steps, each incremental step of the digitizer must correspond to $V_s/2^n$ volts. However, because the signal is continuous and sampling is periodic, between any two samples there is a small voltage differential that the sampler cannot account for. Thus, there is a small error, called *quantization error,* which creates *quantization noise.* Clearly, the larger the n, the more the steps and the smaller the quantization error and quantization noise. The average quantization noise power, $S_{n,q}$, into a 1 Ω load and for a quantization step height q is determined to be

$$S_{n,q} = q^2/12$$

If the expected analog input signal is $x(t)$ and the expected output signal is $y(t)$, then the expected average power error or noise is $E[y(t) - x(t)^2]$ and the expected average input power is $E[x(t)^2]$. Thus, a signal-to-quantization-noise ratio, or signal-to-distortion ratio, or simply signal-to-noise ratio can be formulated, and in order to discriminate it from other SNR sources, we place the suffix q (SNR_q):

$$SNR_q = E[x(t)^2]/E[y(t) - x(t)^2]$$

In the case of uniform quantization (all quantization steps are the same), the quantization error is independent of the sample amplitude. If V_{rms} is the rms value of the input, the SNR_q is determined (in dB) as

$$SNR_q = 10 \log_{10}[V_{rms}^2/(q^2/12)] = 10.8 + 20 \log_{10}[V_{rms}/q]$$

Having explained the significance of power in the bit of a signal, E_b, and the noise power in the same spectral range, N_0, then the ratio bit power to noise power, E_b/N_0, is defined. In addition to this ratio, the carrier power, C, at the receiver is an important parameter. Thus, if the power of noise and the ratio of carrier to noise, C/N, are known, then C is calculated from

$$C = (C/N)N$$

or in dB units:

$$C \text{ (dB)} = C/N \text{ (dB)} + N \text{ (dBm)}$$

where N is the net power noise, including quantization noise. In optical transmission systems, N is calculated by summing all known noise contributors such as ASE, Boltzmann noise ($N_B = kTB$), and so on, including noise figure and noise margin. In the product kTB, k is the Boltzmann's constant, T is the absolute temperature and B is the bandwidth of the passband filter for the channel of interest.

Notice that in the above relationship, the carrier power is, for all practical purposes, the RMS value of the signal power at the receiver. Therefore, there must be a close relationship between C/N and E_b/N_0.

Thus, the CNR ratio is calculated in terms of E_b/N_0, the bit rate, R_b, and the channel bandwidth, B, according to

$$CNR = [E_b/N_0][R_b/B]$$

or in dB units,

$$CNR \text{ (dB)} = E_b/N_0 \text{(dB)} + R_b/B \text{ (dB)}$$

The ratio of bit rate, R_b, to receiver bandwidth, B, is a quantity of interest known as the *power bandwidth ratio* (PBR):

$$PBR = R_b/B \text{ (bits/s/Hz)}$$

Based on the latter,

$$CNR = [E_b/N_0][PBR]$$

Notice that at the receiver, the channel bandwidth B, for spectral matching, is equal to the filter bandwidth. It also follows that a small change in E_b/N_0 causes a large change in CNR, which also causes a large change in BER.

Thus, having calculated the quantities CNR and N, then the carrier power, C, required at the receiver is calculated. From the above, when the bit rate increases, the energy per bit decreases. As the power per bit increases, the interference increases and, thus, the noise content increases. That is, the relationship of signal to noise at the expected quality of signal requires careful characterization of the channel parameters over the optical span. From the aforementioned relationships, the following relationship is also derived:

$$[E_b/N_0] = [CNR]/[PBR]$$

or in dB units,

$$E_b/N_0(\text{dB}) = 10\log(CNR) - 10\log(PBR)$$

In the latter, the ratio CNR is actually the RMS signal power to noise power at the receiver. It also follows that a small change in E_b/N_0 causes a large change in CNR, which causes an equally large change in BER.

7.6 SHANNON'S LIMIT

As noise with respect to signal increases, it becomes obvious that there is a limit beyond which the signal becomes so corrupted that is considered unintelligible. In such a case, the bits in the signal cannot be recovered correctly. Shannon set mathematically the limit of channel capacity (in bits/s), C, in terms of the sampling rate and CNR:

$$C = R_b \log_2\{1 + C/N\}$$

where τ is the interval of sampling rate R_b ($R_b = 1/\tau$). If the sampling rate is equal to the bit period and W is the actual channel bandwidth (in hertz), then,

$$C = R_b \log_2\{1 + [R_b/W][E_b/N]\}$$

In terms of the signal-to-noise ratio (in dB), the latter is expressed as (see also Section 7.13)

$$C = W \log_2[1 + SNR]$$

When the bandwidth equals the bit rate, the bandwidth is referred to as *normalized bandwidth*. In such a case, Shannon's limit is higher than the actual limit. For example, if for a given BER the ratio is $CNR = 10$ dB, Shannon's limit for E_b/N_0 is 1.6 dB and not -11 dB.

7.7 OPTICAL SIGNAL-TO-NOISE RATIO

The optical signal-to-noise ratio (OSNR) is the ratio of RMS power of signal to RMS power of noise. Therefore, for all practical purposes, OSNR is equal to C/N (or CNR), or

$$OSNR = C/N = [E_b/N_0][R_b/B]$$

The ratio E_b/N_0 in the latter relationship is a function of the modulation characteristics; that is, modulation method, modulation efficiency, and modulation loss. Therefore, this ratio needs to be expressed individually for each modulation case.

OSNR is directly related to the optical bit error rate (OBER) and optical error probability (Ope). However, OBER is not observable until the optical signal is detected by a photodetector and converted to an electrical signal.

When the signal is received and converted to an electrical signal, the OSNR needs to be corrected and a corrected SNR is calculated to reflect the receiver and filter noise and gain. In this case, if the particular modulation energy loss factor is β^2, the filter bandwidth is B, and the total noise is N_T, then the SNR is

$$SNR = \beta^2[E_b/N_T][R_b/B]$$

It is this SNR that is directly related to the bit error rate (BER) or the error probability, P_ε.

7.8 PROBABILITY AND STATISTICS 101

As already discussed, probability and statistics play an important role in estimating optical channel performance and, thus, the quality of signal. Before we continue, we should explain in plain terms what is probability, using some popular examples. To begin with, probability starts off by identifying all possible outcomes. Tossing dice or a coin, two ancient games, are based on the probability that a state has to occur. For example, in tossing a coin there are only two possible outcomes: either "heads" or "tails," or "heads" or "nave," as in ancient Greece one side of their coins had the "head" of a ruler and the other side a ship (nave) to signify the city's naval power (Figure 7.3). What is truly important here is that the "head and tail" or "head and ship" represent two distinctive symbols with probability of occurrence of each symbol equal to 1/2. Dice have six faces numbered from 1 to 6 (Figure 7.4) and, thus, there are six symbols and the probability of occurrence 1/6.

Thus, tossing a coin once, the probability that it will land with the "head" or "tail" up is equal to 1/2 and tossing a die it is 1/6. That is, the probability is $p = 1/$(number of expected outcomes).

Now consider that we have two dice. Since each has six symbols, there are $6 \times 6 = 36$ potential outcomes. However, when we toss two dice simultaneously and we expect the numbers 3 and 5, we do not care which die will come up with a 3 and which with a 5. Thus, from all 36 possible outcomes or combinations, we must subtract the double occurrences such as 3 and 5 or 5 and 3. A simple inspection of this simple case reveals that there are six unique outcomes without double occurrences such as 1 and 1, 2 and 2, 3 and 3, 4 and 4, 5 and 5, and 6 and 6. All others, such as 2 and 3, and 4 and 6 have a double occurrence that needs to be subtracted (such as 3 and 2, 6 and 4, and so on). Thus, the total number of possible outcomes in the case

Figure 7.3. The "head or ship" of the ancients was today's "heads or tails" game of chance. Ancient Greek coins with ships on one side from three different cities (left to right: Kyzikos, Phases, and Magnesia).

of two dice is $6 + (6 \times 6 - 6)/2$. One can then generalize that the total number of outcomes of two concurrencies, each with N symbols, is $N + (N \times N - N)/2$, and the probability that any of the possible outcomes will occur is the inverse of it, $1/[N + N(N-1)/2]$.

Consider that one die is thrown two successive times. Then, the order of outcomes is significant. That is, a 3 followed by a 5 is a different from a 5 followed by a 3. In such case, the probability is $(1/6) \times (1/6) = 1/36$, or in the case of succession of two events, each with an outcome of N, it is $1/N^2$. Now, assume that a die is thrown many times independently (say 1000) and that the frequency of occurrence of each outcome is tallied. Since each outcome has the same probability,

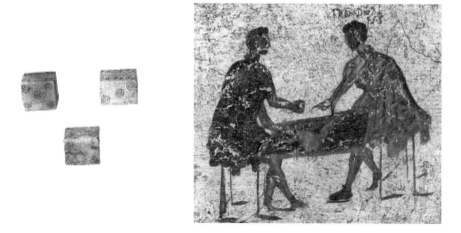

Figure 7.4. Games of chance in antiquity were as popular as they are today, as the three ancient dice and the mural from the Graeco-Roman era attest.

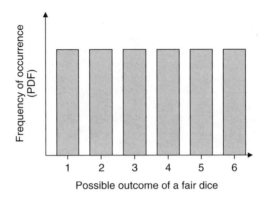

Figure 7.5. Frequency of occurrence of fair dice after many trials.

then after 1000 independent trials each number will have occurred almost equally. If we plot the outcome of this experiment we obtain a flat horizontal line (Figure 7.5). This is the distribution of the occurrences or the probability distribution function (PDF). However, If the die is biased so that a particular number occurs more than the others (such as the number 4), then the PDF is not flat any longer (Figure 7.6).

As another example, consider a pair of fair dice. Since each of the six digits of a single die has a probability of 1/6, any two digits have a probability of 1/36. Thus, if the expected outcome is the sum of any two digits of the thrown pair, then there are three possible combinations to obtain the sum 8 (2 + 6, 3 + 5, and 4 + 4), two combinations to obtain the sum 5 (1 + 4 and 2 + 3) and only one to obtain the sum 2 (1 + 1). Thus, the probability distribution for all possible sums from 2 to 12 is

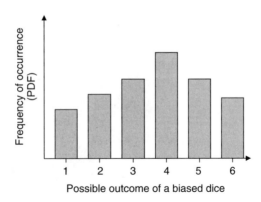

Figure 7.6. Frequency of occurrence of biased dice after many trials.

x_i	2	3	4	5	6	7	8	9	10	11	12	Total
$f(x_i)$	1/36	2/36	3/36	4/36	5/36	6/36	5/36	4/36	3/36	2/36	1/36	36/36 = 1

Plotting these outcomes yields a discrete histogram or probability distribution (Figure 7.7). In nature, there are many phenomena and events that occur in their own idiosyncratic manner following a PDF of particular shapes. The PDF can be discrete if X is a discrete variable (histogram) or continuous if X is a continuous variable (dotted line in Figure 7.7). Notice that in either case, the area under the distribution curve must be equal to one (in the example, it is 36/36 = 1). Thus, the probability that an outcome greater than a value of X occurs is calculated from the area under the distribution from that value on; for example, the probability that the two dice yield a sum a number greater than eight is the area under the PDF from 9 to 12, which is 10/36. This is written as $P(X > 8)$. Similarly, the probability that the sum will be between 4 and 8 eight is 15/36. This is written as $P(4 < X < 8)$. As a last remark, if the probability of success p to toss the number 8 is $p = 5/36$, then the probability of failure q is $q = 1 - p = 1 - 5/36 = 29/36$.

When we deal with random continuous variables X, the probability distribution function is mathematically described by a function $p(X)$ known as the *probability density function* (PDF). Typically, distributions of random variables are named after those who first studied them, or after a major characteristic of the distribution. Thus, we have binomial, Poisson, Gaussian or normal, Fermi–Bose, and so on. The parameters of interest in those distributions are the mean value (μ), which provides the average value; the standard deviation (σ); the variance (var = σ^2), which provides the spread or dispersion of the variable; the moment coefficient of skewness (α_3); and the moment coefficient of kurtosis (α_4). Kurtosis is the degree of narrowness or flatness of the distribution. Here, we outline the three most common distributions; a complete treatment can be found in numerous textbooks on probability and statistics, which can be introductory up to the most advanced level.

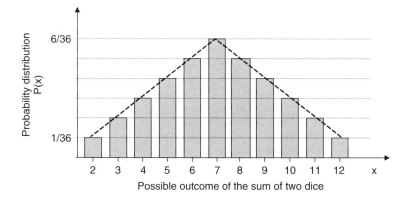

Figure 7.7. Probability distribution of two fair dice.

For our purpose, it suffices to mention certain useful formulas applicable to discrete random variables that will be encountered in this book. Thus, for a random variable with N values, x_i, and corresponding probability of occurrence $f(x_i)$ for each one we have the following:

- Arithmetic mean: $\Sigma x_i/N$, where x_i is a set of N numbers
- Root mean square: $\sqrt{\Sigma x_i^2/N}$, where x_i is a set of N numbers
- Mean or expectation value: $\mu = \Sigma x_i f(x_i)$, summed from 1 to n samples
- Median: in an ordered set of numbers, the median is the middle value of the set
- Mode: the value or values of x_i for which the distribution has maxima
- Variance: $\text{var}(X) = \Sigma(x_i - \mu)^2 f(x_i) = \Sigma x_i^2 f(x_i) - \mu^2 = E(X^2) - \mu^2$
- Standard deviation: $\sigma = \sqrt{\text{var}}$
- rth moment: $m_r = [\Sigma(x_i - \mu)^r]/N$; for $r = 1$, $m_1 = 0$; for $r = 2$, $m_2 = \sigma^2$, and so on
- Skewness: $S \sim 3(\text{mean} - \text{median})/\sigma$
- Moment coefficient of kurtosis: $\alpha_4 = m_4/\sigma^4$

Notice that $f(x_j)$ is the probability of occurrence of each x_i such that $\Sigma f(x_j) = 1$. Some useful properties are:

- $\text{Var}(x + k) = \text{Var}(x)$
- $\text{Var}(kx) = k^2 \text{Var}(x)$
- $\sigma(x + k) = \sigma(x)$
- $\sigma(kx) = |k|\sigma(x)$

where x is a variable and k is a constant.

When mode is less than mean, the distribution is skewed to the right. Similarly, it is skewed to the left if mode is greater than mean. The normal distribution has mean = mode = median.

7.8.1 Binomial Distribution

The probability that an event will occur k times in n trials is

$$P(k) = [n!/(k!(n-k)!)]p^k q^{n-k}$$

Mean:	$\mu = np$
Standard deviation:	$\sigma = \sqrt{npq}$
Variance:	$\sigma^2 = npq$
Skewness:	$\alpha_3 = (q - p)/\sigma$
Kurtosis:	$\alpha_4 = 3 + (1 - 6pq)/\sigma^2$

Where p is the probability of success and q is the probability of failure.

7.8.2 Gaussian and Normal Distributions

These distributions are expressed as

$$P(X) = (1/\sigma\sqrt{2\pi}) \exp\{-\tfrac{1}{2}[(X-\mu)^2/\sigma^2]\}$$

The Gaussian distribution is frequently expressed in the normal form, replacing $(X - \mu)/\sigma$ by the normally distributed variable z. Then, the *normal distribution* is

$$P(X) = (1/\sqrt{2\pi}) \exp\{-\tfrac{1}{2}z^2\}$$

Then, for the normal distribution:

Mean: μ
Mean deviation: $\sigma\sqrt{2/\pi} = 0.7979\sigma$
Standard deviation: σ
Variance: σ^2
Skewness: $\alpha_3 = 0$
Kurtosis: $\alpha_4 = 3$

For the normal distribution, it follows that (Figure 7.8):

One σ on either side of the mean value includes 67% of all cases
Two σs on either side of the mean value includes 95.45% of all cases
Three σs on either side of the mean value includes 99.73% of all cases

7.8.3 Poisson Distribution

A distribution of many natural events is

$$P(X) = \lambda^X e^{-\lambda}/X!$$

where $X = 0, 1, 2, \ldots$, $e = 2.71828 \ldots$, and λ is a given constant.

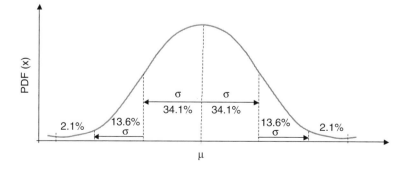

Figure 7.8. Normal distribution and the effect of standard deviation.

A useful expansion in this case is $e^x = 1 + x + x^2/2! + x^3/3! + \ldots$. Then, for the Poisson distribution:

Mean: $\mu = 1$
Standard deviation: $\sigma = \sqrt{\lambda}$
Variance: $\sigma^2 = \lambda$
Skewness: $\alpha_3 = 1/\sigma$
Kurtosis: $\alpha_4 = 3 + 1/\sigma^2$

Note: Unlike the normal distribution, the Poisson distribution changes form as λ changes. For $\lambda = 1$ it is half a normal and for very large λ (>10), it looks like the normal.

7.9 BIT ERROR PROBABILITY

An errored bit is a random process and thus its mathematical treatment is based on stochastic processes and probabilities. In communications, the ratio of detected errored bits, ε, to total bits transmitted, n, is expressed as $P(\varepsilon)$:

$$P(\varepsilon) = \varepsilon/n$$

Typically, the probability of error is estimated as $P'(\varepsilon)$; this is the mean value of the distribution of a random variable and also equal to the reciprocal value of the recurrence time measured in number of bits. However, if the sample of bits is very large, then the estimated probability approximates the actual value.

In communications, an upper limit of error $P'(\varepsilon)$ is specified γ, such as $\gamma = 10^{-N}$, where N may vary from 5 to 16 depending on application, and errored bits are calculated over a period to determine the actual error probability, $P(\varepsilon)$; $P(\varepsilon)$ must be equal or less to $P'(\varepsilon)$. To predict the acceptable error probability, one resorts to statistical methods. Thus, if p is the probability that a bit will be erroneous, and q is the probability that a bit will not be erroneous, such that $p + q = 1$, then the probability that k bit errors will occur in n transmitted bits is expressed by the binomial distribution.

$$P_n(k) = \{n!/(k!(n-k)!)\}p^k q^{n-k}$$

Similarly, the probability that N or fewer errors will occur in n transmitted bits is expressed by

$$P(\varepsilon \leq N) = \Sigma P_n(k) = \Sigma [\{n!/(k!(n-k)!)\}p^k q^{n-k}]$$

where the sum is calculated from $k = 0$ to N, and in the case of N or more errors,

$$P(\varepsilon - N) = 1 - P(\varepsilon \leq N) = \Sigma P_n(k) = \Sigma [\{n!/(k!(n-k)!)\}p^k q^{n-k}]$$

where the sum is calculated from $k = N + 1$ to n.

The above equations are simplified if errors are Poisson random, in which case and for *n* very large (almost infinite), $p^k q^{n-k} = \{(np)^k/k!\}e^{-np}$, and thus

$$\Sigma P_n(k) = \Sigma \{(np)^k/k!\}e^{-np}$$

7.10 BIT ERROR CONTRIBUTORS

In communications, four contributors affect the quality of signal: attenuation, internal and external noise and jitter, and receiver noise, which are manifested as bit errors. For example:

- Attenuation weakens the optical signal power and makes it more vulnerable to noise.
- Bit spreading and sidetones in the propagating optical signal may drift in the next symbol and thus, although two contiguous symbols in the original bit stream might have been "10," now they may appear at the receiver as "11." This is known as *intersymbol interference* (ISI) and pertains only to a single optical channel.
- In DWDM, bits from one channel influence bits in another channel (either because of nonlinear light–matter–light interactions, or because of spectral shift and channel overlapping) and although two contiguous symbols in the original bit stream of a channel might have been "10," now they may appear as "11." This is called *cross talk* and it pertains to more than one optical channel in a fiber.
- In optical communication we do not have the additive noise of electromagnetic interference as we have in electrical transmission. However, optical communication is not immune to noise, as optical amplifiers (doped fibers, Raman, and SOA) spontaneously emit photons in the spectral band of optical channels, and are also subject to other nonlinear mechanisms (scattering, and so on), which are manifested as *optical noise*.
- The receiver, even if the optical signal has been received in its pristine form, adds its own noise (thermal, shot, 1/f).

Subsections 7.10.1 to 7.10.3 list factors affecting the quality of signal and thus the BER at the receiver.

7.10.1 Optical Nonlinearities

- Stimulated Raman scattering (SRS): OChs (at ~1 W) may behave as a pump for longer wavelengths or ASE, degrading S/N ratio of other OChs. Not a well-known controlling mechanism.
- Stimulated Brillouin scattering (SBS): threshold ~5–10 mW (external modulation), ~20–30 mW (direct modulation). Controlled by lowering the signal power level, or making source linewidth wider than Brillouin BW.

- Four-wave mixing (FWM): created sidebands may deplete OCh signal power. Controlled by channel spacing selection (wide or uneven selection).
- Modulation instability (MI): may reduce SNR due to created sidebands (and thus decrease optical power level).
- Self-phase modulation (SPM): optical intensity changes change phase of signal; this broadens the signal spectrum (pulse width). Operating in the anomalous dispersion region, chromatic dispersion and SPM compensate each other. This may result in spontaneous formation of solitons.
- Cross-phase modulation (XPM): interacting adjacent OChs induce phase changes and thus pulse broadening. Controlled by selecting convenient channel spacing.

7.10.2 Polarization

- Polarization-mode dispersion (PMD): randomly changes the polarization state of a pulse, causing pulse broadening. Controlled by polarization scramblers, polarization controllers, or fiber selection.
- Polarization-dependent loss (PDL): due to dichroism of optical components. It affects S/N ratio and Q-value at the receiver. May be controlled with polarization modulation techniques; not well studied.
- Polarization hole burning (PHB): the selective depopulation of excited states due to anisotropic saturation by a polarized saturating signal in EDFA causes noise buildup. Controlled with depolarized signals or polarization scramblers.

7.10.3 Other Factors

- Dispersion properties: OFAs exhibit all fiber properties, including dispersion.
- Noise accumulation: ASE noise is amplified by subsequent OFAs and thus is cumulative; it may exceed signal level (SNR > 1) or surpass the 0/1 discrimination ability of the receiver.
- Temperature variations: some temperature-sensitive components may require component temperature stabilization (e.g., lasers, receivers) to maintain operation within the specified parameters. Some others may operate within a range of temperature (e.g. 0°–70°C) and only environmental conditioning may be required.

The influence of each contributor on the bit error rate is experimentally studied while keeping the others constant and with minimal effect. Figure 7.9 illustrates the impact of chromatic dispersion on BER.

7.11 BIT ERROR RATE

Consider a digital optical signal arriving at a photodetector. Because of all influences on the signal already outlined, the bits in the signal will not have the same

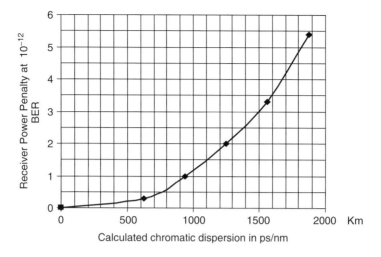

Figure 7.9. Impact of chromatic dispersion on BER.

value they had at the source; that is, some logic "ones" or "zeroes" will have been corrupted beyond recognition. Figure 7.10 illustrates the relationship of bit error rate to received power for a random bit stream (PRBS = $2^{31} - 1$).

To simplify the discussion in this section, we assume that the signal is modulated with on–off keying, NRZ coded, at 100%, and that the photodetector converts the photonic signal to electrical signal without adding its own noise (thermal,

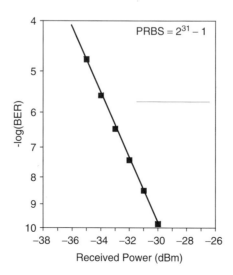

Figure 7.10. Impact of BER on received power (dBm).

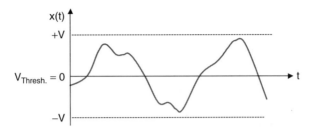

Figure 7.11. Typically, the threshold point in a bipolar signal with noise is set at 0.

shot, 1/f), and, thus, the converted signal provides a true measurement of the optical BER (OBER). OBER, currently, cannot be directly measured in the photonic regime; it can be indirectly measured in the electrical regime and then one can subtract the noise added by the receiver.* Consider now that the converted pulses have a peak voltage $|V_p|$, and that the noise in the signal has a root mean square (RMS) voltage, $\langle Vn^2 \rangle$ (at this point, the photodetector junction noise may be added to V_n). Consider also that the receiver has a decision voltage, b, based on which a determination is made as to whether the received pulse is logic "one" or logic "zero." In symmetric bipolar pulses, the decision voltage is selected such that it is well above the RMS noise for a positive pulse and well below it for a negative pulse; in this case, the threshold is typically set at the zero level, $b = 0$ (Figure 7.11).

In unipolar pulses, the threshold is set at a positive level, the value of which depends on the probability of the noise level in the signal (Figure 7.12).

Now, depending on the actual relative difference of the received amplitudes for V_p and V_n, and on the expected signal voltage, the probability of an errored "one" is $P_{MARK}(V_p - V_n < 0)$. Similarly, if noise is more negative than the signal amplitude, the decision threshold, b, detects the received bit as "zero." Then, the probability for an errored "zero" is $P_{SP}(-V_p + V_n > 0)$. Assuming that the probabilities of an errored "one" and an errored "zero" are equal (in communications, this is a good assumption for a long string of bits), then the total probability that an error to occur, P_ε, is

$$P_\varepsilon = \tfrac{1}{2}P(V_p - V_n < 0) + \tfrac{1}{2}P(-V_p + V_n > 0)$$

When as the probabilities are worked out analytically, the total probability is expressed in terms of the RMS value of the Gaussian standard deviation noise, σ_n, and of the error function erf(·), as

$$P_\varepsilon = \tfrac{1}{2}\{1 - \text{erf}(V_p/(2\sigma_n\sqrt{2}))\}$$

---

*The noise added by the receiver is measured in the electrical regime when either a known test optical signal is applied at the photodetector or no signal at all.

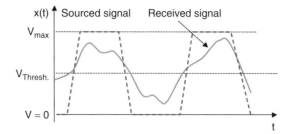

Figure 7.12. The threshold point in a unipolar signal with noise is set at a point that depends on the noise level.

where the function erf(x) = ($2/\sqrt{\pi}$) Int [exp($-y^2$)dy], integrated from 0 to x, and V_p/σ_n is known as the *peak signal to rms noise ratio*.

The latter error probability is also expressed in terms of the *error complimentary function* erf($\cdot$) as:

$$P_\varepsilon = \tfrac{1}{2}\,\mathrm{erfc}(V_p/\sigma_n\sqrt{2})$$

Values of erf($\cdot$) are provided in tables, and erf($\cdot$) is expressed as

$$\mathrm{erfc}(x) = [1/(x\sqrt{2\pi})]e[\exp(-x^2/2)]$$

Figure 7.13 illustrates the dependence of error probability in terms of the peak signal to rms noise ratio.

The standard deviation σ_n is expressed in terms of the number of samples n, the observations x_i, and the mean value x_{mean} of the n observations as

$$\sigma_n = \sqrt{\frac{1}{(n-1)}\,\Sigma(x_i - x_{\mathrm{mean}})^2}$$

The variance, σ_n^2, of a discrete random variable x with mean x_{mean} is by definition

$$\mathrm{Var}(x) = \sigma_n^2 = \Sigma(x_i - x_{\mathrm{mean}})^2 P_x(i)$$

over all samples i of the variable x, where P is the probability function for the discrete case and the mean is

$$x_{\mathrm{mean}} = \Sigma i P_x(i)$$

for all i. In this case, the ratio of peak signal voltage to RMS noise is the SNR.

The probability of error for mark and space is calculated from the probability density functions, the area above the threshold for space, or the area below the threshold for mark (Figure 7.14).

From the above relationships of probability of errored bits, a bit error rate (BER)

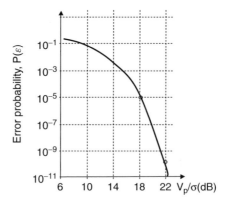

Figure 7.13. The probability of error as a function of peak-signal-to rms-noise ratio (in dB) and for Gaussian noise.

relationship is obtained. Figure 7.13 indicates that the probability of error for "space," $P_{\varepsilon,0}$, is the integration from the threshold point to infinity, and that the probability of error for "mark," $P_{\varepsilon,1}$, is the integration from zero to the threshold point. The total BER then is the sum of both error probabilities for "mark" and "space":

$$BER = P_\varepsilon = \tfrac{1}{2}(P_{\varepsilon,1} + P_{\varepsilon,0}) = \tfrac{1}{2}[\text{Int}\, f_{\varepsilon,1}(x)dx + \text{Int}\, f_{\varepsilon,0}(x)\,dx)]$$

In practice, the BER relationship is expressed in measurable terms of the means (μ_0 and μ_1 for space and mark, respectively), the standard deviations (σ_0 and σ_1 for space and mark, respectively) and decision threshold voltage V_d:

$$BER = \tfrac{1}{2}\,\text{erfc}[(|\mu_1 - V_d|)/\sigma_1\sqrt{2}] + \tfrac{1}{2}\,\text{erfc}[(|\mu_0 - V_d|)/\sigma_0\sqrt{2}]$$

When the above relationship is worked out in terms of the signal-to-noise ratio, then the latter is expressed as

$$BER = \tfrac{1}{2}\,\text{erfc}\,\sqrt{SNR}$$

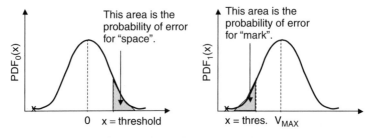

Figure 7.14. Probability of error for "mark" and "space" calculated from corresponding PDF's.

Now, we return our attention to the relationship (for simplicity, restated here)

$$\Sigma P_n(k) = \Sigma \{(np)^k/k!\} e^{-np}$$

The probability that the actual $P(\varepsilon)$ is better than the acceptable set level γ is known as the confidence level, CL, and is defined as (CL is typically expressed in percent, %)

$$CL = P(\varepsilon > N|\gamma) = 1 - \Sigma[\{n!/(k!(n-k)!)\} P^k (1-\gamma)^{n-k}]$$

where the sum is calculated from $k = 0$ to N. The last equation can be solved for n, the number of transmitted bits required to be monitored for errors. Clearly, the number n in the unit of time is directly related to a bit rate. Thus, assuming a confidence level of 99%, a BER threshold of $\gamma = 10^{-10}$, and a bit rate of 2.5 Gb/s,* then the required number n to detect a single error is 6.64×10^{10}.

Thus, if the *probability* of a single bit being in error within the unit of time or within an equivalent block of bits is p, then the probability of two errors occurring independently within the same block is p^2, of three errors it is p^3, and so on.

The BER is directly proportional to the error probability p for a single error, p^2 for two errors, and so on. Thus, for a poor BER = 10^{-3}, what this means is that:

- A single error is *most likely* to occur in 1,000 bits in the unit of time
- A double error is 1,000 times *more unlikely* to occur than a single bit error
- A triple error is a 1,000,000 times *more unlikely* to occur than a single bit error, and so on

As consequence, a triple error is one million times *more unlikely* to occur than a single bit error; this implies an almost never occurring event and it may be a reasonable assumption at very low bit rates, such as Kbps.

7.12 OPTICAL SIGNAL-TO-NOISE RATIO

We have talked about signal-to-noise ratio and (in optical communication) about the *optical signal-to-noise ratio* (OSNR), as a measure of the optical noise in the optical signal. Signal-to-noise ratio is defined as the ratio of available signal power, P_{so}, to the available noise, P_{no}, at the output of a transmitter (laser):

$$SNR = P_{so}/P_{no}$$

SNR is measured in dB, and thus it may be rewritten as

$$SNR \text{ (dB)} = P_{so} \text{ (dBm)} - P_{no} \text{ (dBm)}$$

*Notice that here we state the bit rate, as bit error rate has no meaning without it.

If P_{si} and P_{ni} are the available signal and noise power at the input of the receiver, then the noise figure (NF) is expressed in terms of the OSNR as

$$NF = \frac{P_{si}/P_{ni}}{P_{so}/P_{no}} = \frac{P_{si}/kT_0 df}{OSNR}$$

The SNR itself is a complex parameter, as noise is the compounded effect of all noise, jitter, attenuation, and other degradations that affect the form of the signal.

The peak signal-to-rms-noise ratio V_p/σ_n is a parameter in the error expression

$$P_\varepsilon = \tfrac{1}{2}\{1 - \mathrm{erf}[V_p/(2\sigma_n\sqrt{2})]\}$$

A probability of error 10^{-10} means one expected error in 10,000,000,000 bits transmitted. A $P_\varepsilon = 10^{-10}$ corresponds to a peak signal-to-rms-noise ratio V_p/σ_n of 22 dB. Now, assume that the rms noise increases so that the ratio V_p/σ_n decreases by only 3 dB. 18 dB corresponds to a single error probability of 10^{-5}, and, thus, an error is expected in every 100,000 bits transmitted, a dramatic degradation in signal quality at the receiver for such a small signal degradation.

In addition to Gaussian noise, there are non-Gaussian spectral distributions, such as Poisson and so on. Therefore, the probability for error differs and it depends on the particular noise distribution case. An empirical formula that we have used to calculate the BER of optical signals in single-mode fiber, given an OSNR value in dB, is

$$\log_{10} BER = 1.7 - 1.45\ (OSNR)$$

For example, assume that $OSNR = 14.5$ dB, then $\log_{10} BER = 10.3$ and $BER = 10^{-10}$.

7.13 CHANNEL CAPACITY AND SNR

We have discussed one view of the Shannon maximum capacity of a communication channel in Section 6. Here we provide an in-depth examination.

If the highest frequency (or bandwidth) of an analog signal is B Hz, the signal is sampled at the Nyquist rate $2B$ and each analog sample is converted to M possible amplitude steps. Parenthetically, when a time-dependent analog signal is Fourier transformed, the frequency spectrum of the signal determines a maximum frequency. Typically, the maximum frequency is the upper cutoff frequency of the passband filter through which the analog signal is frequency limited (or filtered from high-frequency noise) (Figure 7.15).

Assuming binary conversion, M steps correspond to n-bit-long code words as n is calculated from $M = 2^n$ and $n = \log_2 M$ (Figure 7.16). If $V_{p\text{-}p}$ is the maximum expected peak-to-peak voltage swing, then the quantization step D is

$$D = V_{p\text{-}p}/M = V_{p\text{-}p}/2^n$$

If the length of the *n*-bit code is increased by one bit from *n* to *n* + 1, the number of steps *M* is doubled, the step-size is cut in half, the resolution is improved, and the *quantization noise improvement* in this case is 20 log$_2$ = 6 dB.

Based on the above, the channel capacity or bit rate of information is

$$C = 2B \log_2 M = 2nB \text{ bits/s}$$

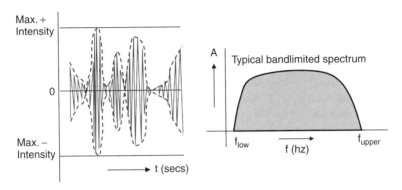

Figure 7.15. Analog signal in the time domain and its frequency spectrum or bandwidth after it passes through a passband filter.

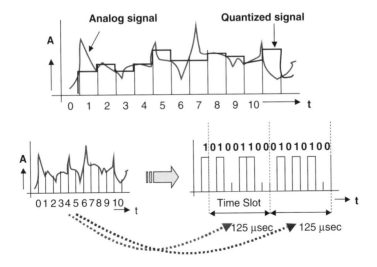

Figure 7.16. The analog signal is sampled (top). Samples are coded to a binary bit stream (bottom). The digital bit stream is decoded to an analog quantized signal. The difference of the two yields a quantization error (top) or noise.

For example, in telephony, $n = 8$, $B = 4$ kHz (the upper cutoff frequency) and the Nyquist sampling rate is $2B = 8$ ksamples/sec, from which the DS0 rate is derived ($8 \times 8000 = 64$ kbits/s), or $2nB$.

Consider now the more general case for which the M levels, instead of binary, are coded into base-m arithmetic.* That is, instead of two levels, we know have m symbols of n bits, or $M = m^n$. Thus,

$$C = 2nB \log_2 m = W \log_2 m \text{ bits/s}$$

where $W = nB$ is the transmission bandwidth. If we assume that transmission is bipolar NRZ, and that the pulse amplitude is $\pm A/2$, then the average signal power is

$$S = (2/m)\{(A/2)^2 + (3A/2)^2 + \ldots + [(m-1)a/2]^2\} = A^2(m^2 - 1)/12$$

Solving for m and replacing it in the channel capacity relation, one obtains

$$C = W \log_2(1 + 12S/A^2) \text{ bits/s}$$

The actual selection of the amplitude A depends on the noise variance σ^2 and the error probability P_e. In general, if $A = K\sigma$, then

$$C = W \log_2[1 + (12/K^2)(S/N)] \text{ bits/s}$$

That is, the maximum bit rate, C, transmitted over a channel is limited by the channel bandwidth W and the S/N ratio. This stems from the interplay of many consequences. For example, the higher the bitrate, the narrower the pulse, the shorter the bit period, the less average power per bit (assuming constant peak power level), the higher the frequency content (per Fourier transform), the higher the quantizing power spectrum, the higher the power noise, and the higher the probability for bit error.

The probability for a bit error, due to presence of noise (of any type), in n transmitted bits approaches certainty as n becomes very large. As noise increases, the probability for a bit error increases rapidly (that is, an error will occur in fewer than n bits). Thus, the fundamental question "what is the maximum transmission bit rate, if the channel bandwidth and the relationship of signal to noise are known?" was answered by Shannon:

$$C = W \log_2(1 + S/N)$$

At this point, certain observations can be made about Shannon's channel capacity:

- Shannon's theorem applies to the channel and not to the data source.
- For a given channel, when the relationship $R \leq C$ holds, where R is the rate across the channel with capacity C, then there is error-free transmission.

---

*Binary arithmetic uses two symbols, and thus for an n-bit code word there are 2^n combinations or levels. A 10-base arithmetic uses 10 symbols [0; 9] and an n digit codeword has 10^n combinations or levels. The same arguments are used in m-ary arithmetic.

- For a binary channel, the channel capacity is described in terms of the probability p that the signal through it is without error, $C = 1 + p \log_2 p + (1 - p) \log_2(1 - p)$. If $p = 0$, then $C = 1$ and if $p = 0.5$ then $C = 0$.
- For a binary channel with bandwidth B, $2B$ transitions can be transmitted. Thus, the maximum capacity is $2BC$ bits/s and the capacity of the binary channel is $C_{max} = 2B[1 + p \log_2 p + (1-p)\log_2(1-p)]$. The maximum rate W is then $2B$. The relationship of W, B, and the actual datarate D is $W > B > D$

7.14 THE QUALITY OR Q-FACTOR

The quality factor, or Q-factor, is a parameter that is closely related to the quality of the transmitted optical signal in terms of signal-to-noise ratio and bit error rate. Mathematically, the Q-factor is an intermediate step in the derivation of the bit error rate and signal-to-noise ratio (SNR). It is defined as the ratio, after observing a large number of 1/0 symbols, of the difference of means for 1 and for 0 symbols ($\mu_1 - \mu_0$) and the difference of standard deviations for 1 and for 0 symbols ($\sigma_1 - \sigma_0$):

$$Q = (|\mu_1 - \mu_0|)/(|\sigma_1 - \sigma_0|)$$

In traditional communications, it has been established that in order to achieve a probability of one errored symbol in 10^{10} transmitted, a SNR greater than 15 dB is required. In fact, because with small increases in SNR the probability of error decreases rapidly, particularly above 15 dB, the region above SNR = 15 dB is known as *the cliff*.

The aforementioned relationship for Q is assumed to have a Gaussian distribution and equal error probabilities for both 1 and 0 symbols. It has also been assumed that the received signal is bipolar and that the threshold voltage is $Vd = 0$).

When we consider optical signals, the modulated signal varies at best between a zero and a positive power level. Thus, the threshold, Vd, is not set to zero but at some positive value. Then, the probabilities of bit error for the amplitude-shift keying (ASK) modulation method (coherent or synchronous) and for on–off keying (OOK) are given by:

$$\text{ASK (Coherent): } Pe = \tfrac{1}{2} \operatorname{erfc} \sqrt{S/4N}$$

and

$$\text{OOK: } Pe = \tfrac{1}{2}\sqrt{S/N}$$

where the function erfc is the complimentary error function, the value of which is obtained from mathematical tables. Table 7.1 lists BER and corresponding OSNR approximate values and Figure 7.17 provides a relative graph of the BER performance of two modulation methods, OOK and PSK.

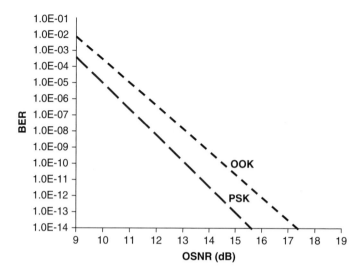

Figure 7.17. Approximate relationship of BER versus OSNR in single-mode fiber for two modulation methods, OOK and PSK.

Example 1

A random discrete variable x has the following probability function: $P_x(i) = 1/10$ for $i = 2$, $P_x(i) = 2/10$ for $i = 3$, $P_x(i) = 4/10$ for $i = 4$, $P_x(i) = 2/10$ for $i = 5$, and $P_x(i) = 1/10$ for $i = 6$. Calculate: (A) the mean value, (B) the standard deviation.

(A) From the relationship $x_{\text{mean}} = \Sigma i P_x(i)$ for all i, the mean value is calculated:

$$x_{\text{mean}} = (1/10)(2) + (2/10)(3) + (4/10)(4) + (2/10)(5) + (1/10)(6) = (40/10) = 4$$

(B) From the relationship $\sigma = \sqrt{1/(n-1) \Sigma(x_i - x_{\text{mean}})^2}$, the standard deviation is calculated:

$$\sigma = \sqrt{1/5[\{(1/10)-4\}^2 + \{(2/10)-4\}^2 + \{(4/10)-4\}^2 + \{(2/10)-4\}^2 + \{(1/10)-4\}^2]}$$
$$= 1/5 \sqrt{15.21 + 14.44 + 12.96 + 14.44 + 15.21} = \sqrt{14.45} = 3.80$$

Table 7.1. BER and corresponding SNR values (for OOK)

BER—Pe	OSNR (dB)
10^{-10}	19.4
10^{-9}	18.6
10^{-8}	18.0
10^{-7}	17.3
10^{-6}	16.4
10^{-5}	15.3

Example 2
Calculate the probability error for the ASK (coherent) case if the S/N ratio is 18 dB. The S/N power ratio is calculated from $18(dB) = 10 \log x$ as $x = 63.36$.

Based on this, the probability error is calculated:

$$P_e = \tfrac{1}{2} \text{ erfc } \sqrt{63.36/4} = \tfrac{1}{2} \text{ erfc } \sqrt{15.84} = \tfrac{1}{2} \text{ erfc } (3.98) = 1 - 2(1.8 \times 10^{-8}) = 9 \times 10^{-9}$$

7.15 BIT ERROR MONITORING

To monitor signal quality, one or more techniques may be used: *termination, sampling, spectral monitoring,* or *indirect.*

The termination technique consists of error-detecting codes (EDCs) that have been incorporated into the information bit stream of the source. At the receiver, an EDC code checks the incoming bit stream bit by bit and finds erroneous bits, which are counted to deduce the actual BER. However, EDC codes have limited detecting and correcting capability.

The sampling method requires fast discrete sampling circuitry, a programmable synchronizer to sample the received signal, a fast analog-to-digital converter, a signal processor, and a signal analyzer that consists, at a minimum, of a demultiplexer, a very-low-noise detector with programmable sensitivity, and an algorithmic discrete arithmetic unit.

Spectral monitoring consists of noise level measurements and spectral analysis, and is capable of performing fast Fourier transforms.

If no direct signal monitoring is performed (noise, distortion, power spectra, etc), then the quality of signal depends on the quality of system design. Thus, signal quality is grossly computed from indirect information (in a manner similar to that of legacy systems) such as system alarms (loss of frame, loss of synchronization, and so on).

7.16 BER AND DATA PATTERNING

The previous estimates have been based on a significant assumption: there is equal probability for "mark" and "space," or $p_1 = p_0 = 0.5$. However, there are cases in which this is not so, due to what is known as the patterning effect. Patterning effects appear due to fiber nonlinearities and dispersion when they cause strong intersymbol interference; in such cases, errors may favor one symbol over the other. As a result, a specific bit in the bit stream is affected by the bit that precedes it and the bit that follows it. Since there are three concatenated bits, the possible patterns are 000, 001, 010, 011, 100, 101, 110, and 111.

In the case of patterning, the analysis proceeds as previously by calculating the error probabilities, standard deviation, and mean values for both "mark" and "space," although in this case the error probability distributions are not the same due to the occurrence of patterns in data. When patterning is considered, a correction of the Q-factor and BER is obtained. At 1.5 GHz the standard method provides $Q = 15.4$ and the patterning method $Q = 15.74$ (see reference 27).

A simpler way of estimating patterning is by considering the case of a coin that is biased. That is, the probability of tossing heads ("mark") is not equal to the probability of tossing tails ("space"). In this case, assuming normal distributions for both heads and tails, but for different probabilities, say $p_1 = 0.6$ for heads and $p_0 = 0.4$ for tails, the error is calculated from the overlapping regions (see Figure 7.21), although in this case the two distribution have different degrees of kurtosis.

7.17 BER AND EYE DIAGRAM

One quick and qualitative measure of the quality and integrity of the received signal is a superposition of bit periods on an oscilloscope. This superposition yields what is known an "eye diagram" (Figure 7.18). Clearly, this can be achieved only in the electronic regime.

The eye diagram reveals on the aggregate all characteristics of the signal, such as jitter, rise and fall time, ringing, side-tones, asymmetry (skew and kurtosis), overshoot and undershoot, and amplitude tilt and variability (Figure 7.19).

If the signal has little noise and its amplitude is clearly recognized as a "one" or "zero," then the superposition provides an "open eye." If noise and distortion are excessive, then the eye appears corrupted and "fuzzy," (Figure 7.20) (as already discussed in Chapter 5; see Figure 5.3).

Based on the expected error probabilities for logic "1" and for logic "0," the threshold (or decision) level V_d is determined (Figure 7.21).

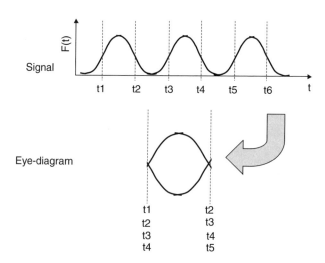

Figure 7.18. Eye diagram construction.

7.17 BER AND EYE DIAGRAM

In addition to the threshold voltage level, a sampling point that lies within the bit period must also be determined. Again, this is determined from the expected shape degradation (skew, shift, sidetones, dispersion) and also jitter. Typically, the sampling point within the period is about the $0.5\,T$ to $0.6\,T$ point (Figure 7.22).

The *bit error rate* (BER) and the opening of the eye are closely related. However, the eye opening is the observed electrical signal and thus, as is known from dig-

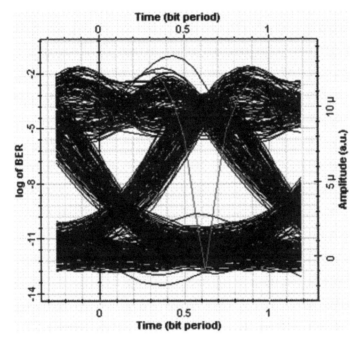

Figure 7.19. An actual eye diagram that clearly reveals rise/fall time, ringing, overshoot, undershoot, and jitter.

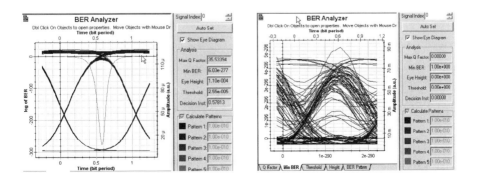

Figure 7.20. Actual eye diagrams. Open (left) and corrupted (right).

242   PROBABILITY THEORY OF BIT ERROR RATE

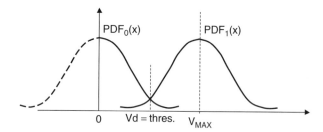

Figure 7.21. The threshold level is determined from the lowest probability for error for "mark" and "space" error probability distributions. This is where the tails of the two distribution overlap. Notice that the probability for "zero" has been truncated as there is no negative optical signal.

ital sampling, there are two current thresholds, one to determine the logic 1 and another one to determine the logic 0. Thus, assuming no jitter, at the half-period of the bit rate there is a minimum current threshold (through the unit load) for a logic 1, $I_{1,\min}$, and a maximum current threshold (through the unit load) for a logic 0, $I_{0,\max}$. Then, the eye opening is defined as

$$E_{\text{eye}} = I_{1,\min} - I_{0,\max}$$

Now, noise and jitter are random effects, most likely with a Gaussian distribution. If the standard deviation for a logic "1" bit (or mark) is σ_1 and for a logic "0" bit (or

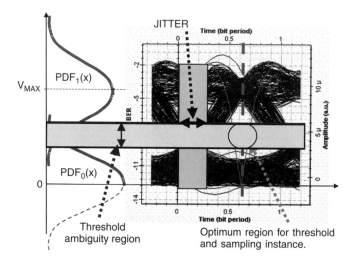

Figure 7.22. Anatomy of an eye diagram showing jitter and noise. The threshold level is determined from the lowest error probability for "mark" and for "space." This is where the tails of the two distribution overlap.

space) it is σ_0, their corresponding average values are μ_1 and μ_0, the quality factor, Q (which is a measure of the signal-to-noise ratio), is defined as

$$Q = E_{eye}/\sqrt{|\sigma_0^2 - \sigma_1^2|} = |\mu_1 - \mu_0|/(\sigma_1 + \sigma_0)$$

Then, the BER is defined as

$$BER = \tfrac{1}{2}\operatorname{erfc}(Q/\sqrt{2})$$

If noise is Gaussian, the BER is also expressed as:

$$BER = [\exp(-Q^2/2)]/(Q\sqrt{2\pi})$$

Based on this, a Q value of 6.1 corresponds to $BER = 10^{-9}$, and a Q value of 7.2 to $BER = 10^{-12}$.

Some relationships that estimate various parameters of the eye-diagram are:

Eye opening: $E_{eye} = (\mu_1 - \mu_0)$
Eye height: $E_{height} = (\mu_1 - 3\sigma_1^2) - (-\mu_0 + \sigma_0^2)$
Extinction ratio*: $R_{ext} = \mu_1/\mu_0$

In addition, if the expected eye opening is E_{eye} and the measured average eye opening is E_{avg}, then a quantitative penalty ratio is defined as

$$\text{penalty} = E_{eye}/2E_{avg}$$

The aforementioned analysis has assumed no timing degradations. Timing degradations include jitter and static decision-time misalignment. If we include these degradation, an adjustment in the sampling point in the bit period T must be made, displacing it from $T/2$ by $\Delta\tau$ (Figure 7.23). This also causes an adjustment in the two threshold values (because of the new sampling point); the eye opening in this case is denoted as $E(\tau)$, and the $I_{1,min}$ and $I_{0,max}$ are determined at the new sampling point.

Figure 7.24 defines the sum of all contributors responsible for amplitude degradation, the maximum peak-to-peak eye, E_{max}, and the eye opening, E_{eye}. Then, the signal degradation is expressed in terms of the maximum and open eye as

$$\Delta S/N = 20 \log [E_{max}/E_{eye}]$$

7.18 EYE-PATTERN MASK

To control the optical channel and the expected signal at the receiver, it is important that the signal at the transmitter conform to a specific quality, which may also be

---

*Per ITU-T G.693, the extinction ratio is defined as $10 \log_{10}(A/B)$, where A and B are the average optical power level at the center of a "mark" and "space" symbol, respectively.

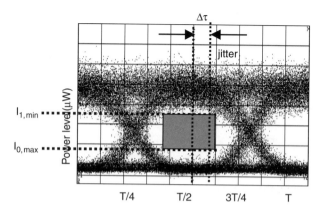

Figure 7.23. Threshold levels (power and jitter) mapped on an eye diagram.

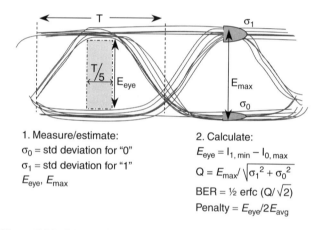

1. Measure/estimate:
σ_0 = std deviation for "0"
σ_1 = std deviation for "1"
E_{eye}, E_{max}

2. Calculate:
$E_{eye} = I_{1,min} - I_{0,max}$
$Q = E_{max}/\sqrt{\sigma_1^2 + \sigma_0^2}$
BER = ½ erfc $(Q/\sqrt{2})$
Penalty = $E_{eye}/2E_{avg}$

Figure 7.24. BER and penalty estimation based on an eye diagram.

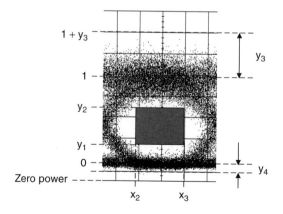

Figure 7.25. Threshold levels (power and jitter) mapped on an eye diagram.

Table 7.2. Parameters of the eye-pattern mask for NRZ 10 Gbit/s in the C-band (per ITU-T G.693)*

	NRZ, 10 Gbit/s, 1550 nm region
x_3-x_2	$0.2\,T$
y_1	$0.25\,\mu_1$
y_2	$0.75\,\mu_1$
y_3	$0.25\,\mu_1$
y_4	$0.25\,\mu_1$

*T is the period of the NRZ pulse and μ_1 is the mean value for "mark" distribution ($\mu_0 = 0$).

provided by the eye diagram at the source. ITU-T G. 693 provides an eye-pattern mask on which specific parameters and limits are identified (for 10 Gbit/s and 40 Gbit/s optical bitrates; (Figure 7.25). Some of these parameters for the case NRZ 10 Gbit/s for the C-band region are given in Table 7.2.

In addition, other standards, such as the GR-253-CORE SONET standard, that specify the physical layer also provide similar eye-pattern masks.

REFERENCES

1. A. Papoulis, *Probability, Random Variables, and Stochastic Processes,* McGraw-Hill, 1984.
2. D. Middleton, *An Introduction to Statistical Communication Theory,* IEEE Press, 1996.
3. M. Schwartz, *Information, Transmission, Modulation, and Noise,* McGraw-Hill, 1970.
4. Abramovitch and Stegun, *Handbook of Mathematical Functions,* National Bureau of Standards, 1964.
5. D. Derickson, Fiber Optic Test and Measurement, Prentice Hall, 1998.
6. S.V. Kartalopoulos, *DWDM: Networks, Devices and Technology,* Wiley/IEEE-Press, 2003.
7. S.V. Kartalopoulos, *Fault Detectability DWDM: Technology Data in a Rainbow,* Wiley/IEEE-Press, 2000.
8. T. Wildi, *Units and Conversion Charts,* 2nd ed., IEEE Press, 1995.
9. M. C. Jeruchim, P. Balaban, and K.S. Shanmugan, *Simulation of Communications Systems,* Plenum Press, 1992.
10. R.J. Baker, *CMOS: Mixed Signal Circuit Design,* Wiley/IEEE-Press, 2002.
11. R. Marz, *Integrated Optics: Design and Modeling,* Artech House, 1995.
12. E.N. Gilbert, "Capacity of a Burst Noise Channel," *Bell System Technical Journal, 39,* 1253–1265, 1960.
13. M.C. Jeruchim, "Techniques for Estimating the Bit Error Rate in the Simulation of Digital Systems," *IEEE Journal on Selected Areas in Communications, SAC-2,* 1, 153–170, 1984.
14. V. Ramaswami and L.J. Wang, "Analysis of the Link Error Monitoring Protocols in the Common Channel Signalling Network," *IEEE/Association for Computing Machinery Transactions on Networking, 1,* 1, 31–47, 1993.

15. N.L. Kanal and A.R.K. Sastry, "Models for Channels With Memory and Their Applications to Error Control," *Proceedings of IEEE, 66,* 7, 724–744, 1978.
16. M.D. Knowles and A.I. Drukarev, "Bit Error Rate Estimation for Channels with Memory," *IEEE Transactions on Communications, 36,* 6, 767–769, 1988.
17. C.E. Shannon, "A Mathematical Theory of Communications, Part A," *Bell Systems Technology Journal, 27,* 379–423, 1948.
18. C.E. Shannon, "A Mathematical Theory of Communications, Part B," *Bell Systems Technology Journal, 27,* 623–656, 1948.
19. S.R. Nagle, "Optical Fiber—The Expanding Medium," *IEEE Circuits and Devices Magazine, 5,* 2, 36–45, March 1989.
20. R.H. Stolen, "Non-Linear Properties of Optical Fibers," in *Optical Fiber Telecommunications,* S.E. Miller and G. Chynoweth (Eds.), Academic Press, 1979.
21. N. Gisin, "Statistics of Polarization Dependent Losses," *Optics Communications, 114,* 1995, 399–405.
22. C.R.S. Fludger, V. Handerek, and R.J. Mears, "Fundamental Noise Limits in Broadband Raman Amplifiers," *Technical Digest OFC 2001,* paper MA5-1, March 2001.
23. P.K. Pepeljugoski and K.Y. Lau, "Interfoerometric Noise Reduction in Fiber-Optic Links by Superposition of High Frequency Modulation," *Journal Lightwave Technology, 10,* 7, 957–963, June 1992.
24. I.T. Monroy, E. Tangdiongga, and H. de Waardt, "On the Distribution and Performance Implications of Filtered Interfoerometric Crosstalk in Optical WDM Networks," *Journal Lightwave Technology, 17,* 6, 989–997, June 1999.
25. I.T. Monroy, E. Tangdiongga, and H. de Waardt, "Interfoerometric-Crosstalk Reduction by Phase Scrambling," *Journal Lightwave Technology, 18,* 5, 637–646, May 2000.
26. F. Matera and M. Settembre, "Role of Q-factor and of Time Jitter in the Performance Evaluation of Optically Amplified Transmission Systems," *IEEE Journal of Selected Topics in Quantum Electronics, 6,* 2, 308–316, 2000.
27. C. J. Anderson and J.A. Lyle, "Technique for Evaluating System Performance Using Q in Numerical Simulations Exhibiting Intersymbol Interference," *Electronics Letters, 30,* 1, 71–72, 1994.
28. S.V. Kartalopoulos, "Factors Affecting the Signal Quality and Eye-Diagram Method For The Estimation of BER and SNR in Optical Data Transmission," *Proceedings of the International Conference on Information Technology, ITCC-04,* Las Vegas, April 5–7, pp. 615–619, 2004.
29. P.A. Humblet and M. Azizoglu, "On the Bit Error Rate of Lightwave Systems with Optical Amplifiers," *Journal of Lightwave Technology, 9,* 11, 1576–1582, 1991.
30. E. Forestieri, "Evaluating the Error Probability in Lightwave Systems with Chromatic Dispersion, Arbitrary Pulse Shape and Pre- and Post-Detection Filtering," *Journal of Lightwave Technology, 18,* 11, 1493–1503, 2000.
31. M.R.N. Ribeiro, H. Waldman, J. Klein, and L. de Souza Mendes, "Error-Rate Patterns for the Modeling of Optically Amplified Transmission Systems," *IEEE Journal on Selected Areas in Communications, 15,* 4, 707–716, 1997.
32. E.G. Shapiro, M.P. Fedoruk, and S.K. Turitsyn, "Numerical Estimate of BER in Optical Systems with Strong Patterning Effects," *Electronics Letters, 37,* 3, 304–306, 1993.
33. D. Marcuse, "Derivation of Analytical Expressions for the Bit-Error Probability in Lightwave Systems with Optical Amplifiers," *Journal of Lightwave Technology, 8,* 12, 1816–1823, 1990.

34. Z. Zalevsky and V. Eckhouse, "In-Channel OSNR and BER Measurement Using Temporal Superresolution Via Dyanic Range Conversion," *Journal of Lightwave Technology, 21,* 11, 2734–2738, November 2003.
35. S.V. Kartalopoulos, "Apparatus and method for optical pattern detection," U.S. patent 6,617,566, issued 9/9/2003.
36. J.C. Cartledge and G.S. Burtley, "The Effect of Laser Chirping on Lightwave System Performance," *Journal Lightwave Technology, 7,* 3, 568–573, 1989.
37. J.J. O'Reilly, and J.R.F. da Rocha, "Improved Error Probability Evaluation Methods for Direct Detection Optical Communication Systems," *IEEE Transactions on Information Theory, IT-33,* 6, 839–848, 1987.
38. R. Dogliotti, A. Luvison, and G. Pirani, "Error Probability in Optical Fiber Transmission Systems," *IEEE Transactions Information Theory, IT-25,* 170–178, March 1979.
39. W. Hauk, F. Bross, and M. Ottka, "The Calculation of Error Rates for Optical Fiber Systems," *IEEE Transactions Communications, COM-26,* 1119–1126, July 1978.
40. P. Balaban, "Statistical Evaluation of the Error Rate of the Fiberguide Repeater Using Importance Sampling," *Bell Systems Technology Journal,* 745–766, July–August 1976.
41. G.L. Gariolato, "Error Probability in Digital Fiber Optic Communication Systems," *IEEE Transactions Information Theory, IT-24,* 213–221, 1978.
42. R. Lugannani, "Intersymbol Interference and Probability of Error in Digital Systems," *IEEE Transactions Information Theory, IT-15,* 682–688, 1969.
43. G. Bendelli, C. Cavazzoni, R. Giraldi, and R. Lano, "Optical Performance Monitoring Techniques," *Proceedings of the European Conference on Optical Communication, ECOC 2000, 4,* 113–116, 2000.
44. H. Nishimoto et al., New Method of Analyzing Eye Patterns and its Application to High-Speed Optical Transmission System," *Journal Lightwave Technology, 6,* 5, 678–685, 1988.
45. K. Hinton and T. Stephens, "Modeling High-Speed Optical Transmission Systems," *IEEE Journal on Selected Areas in Communications, 11,* 4, 380–392, 1993.
46. A.J. Lowery, Computer-Aided Photonics Design," *IEEE Spectrum, 34,* 4, April 1997, 26–31.
47. M. Kawasaki, "Silica Waveguides on Silicon and Their Application to Integrated Optic Components," *Optics Quantum Electronics, 22,* 1990, 391–416.

STANDARDS

1. ANSI/IEEE 812-1984, *Definition of Terms Relating to Fiber Optics,* 1984.
2. Bellcore (currently Telcordia) GR-253-CORE, *Synchronous Optical Network (SONET) Transport Systems: Common Generic Criteria,* 1995.
3. CCITT Recommendation O.151, *Error Performance Measuring Equipment Operating at the Primary Rate and Above,* 1992.
4. CCITT Recommendation O.153, *Basic Parameters for the Measurement of Error Performance at Bit Rates Below the Primary Rate,* 1992.
5. ITU-T Recommendation G.650, *Definition and Test Methods for the Relevant Parameters of Single-Mode Fibres,* 1997.

6. ITU-T Recommendation G.652, *Characteristics of a Single-Mode Optical Fiber Cable*, 1997.
7. ITU-T Recommendation G.653, *Characteristics of a Dispersion-Shifted Single-Mode Optical Fibre Cable*, 1997.
8. ITU-T Recommendation G.654, *Characteristics of a Cut-Off Shifted Single-Mode Optical Fibre Cable*, 1997.
9. ITU-T Recommendation G.655, *Characteristics of a Non-Zero-Dispersion Shifted Single-Mode Optical Fibre Cable*, 1996.
10. ITU-T Recommendation G.661, *Definition and Test Methods for the Relevant Generic Parameters of Optical Fiber Amplifiers*, November 1996.
11. ITU-T Recommendation G.671, *Transmission Characteristics of Passive Optical Components*, 2001.
12. ITU-T Recommendation G.691, *Optical Interfaces for Single Channel STM-64, STM-256 and Other SDH Systems with Optical Amplifiers*, 10/2000.
13. ITU-T Recommendation G.692, *Optical Interfaces for Multichannel Systems with Optical Amplifiers*, 10/1998.
14. ITU-T Recommendation G.693, *Optical Interfaces for Intra-Office-Systems*, 11/2001.
15. ITU-T Recommendation G.826, *Error Performance Parameters and Objectives for International, Constant Bit Rate Digital Paths at or Above the Primary Rate*, 1993.
16. ITU-T Recommendation G.828, *Error Performance Parameters and Objectives for International, Constant Bit Rate Synchronous Digital Paths*, 3/2000.
17. ITU-T Recommendation G.829, *Error Performance Parameters Events for SDH Multiplex and Regenerator Sections*, 3/2000.
18. ITU-T G.955, *Digital Line Systems Based on the 1544 kbit/s and the 2048 kbit/s Hierarchy on Optical Fibre Cables*, 1996.
19. ITU-T G.957, *Optical Interfaces for Equipments and Systems Relating to the Synchronous Digital Hierarchy*, 1999.
20. ITU-T Recommendation L.41, *Maintenance Wavelength on Fibres Carrying Signals*, May 2000.
21. ITU-T Recommendation O.150, *General Requirements for Instrumentation for Performance Measurements on Digital Transmission Equipment*, 5/1996.
22. Telcordia (previously Bellcore), TR-NWT-499, *Transport Systems Generic Requirements (TSGR): Common Requirements*, issue 5, Dec. 1993.

CHAPTER 8

BER Statistical Measurements

8.1 INTRODUCTION

At the receiver, the quality of the arriving signal is continuously monitored and performance metrics are periodically reported to network management. Performance monitoring has always been critical in digital communications, even if the bit rate or line rate was very low (1.5 Mbit/s). In optical communications, the bit rate has been increased more than three orders of magnitude up to 40 Gbit/s. Accfordingly, the performance metrics have been changed to reflect the increase in bit rate. For example, from an initial 10^{-5} BER to 10^{-12} for modern optical networks, and it is better than 10^{-9} for telephony, better than 10^{-6} for data, and better than 10^{-4} for telemetry and other low-rate data.

With the advent of DWDM technology that supports an aggregate data rate in excess of terabits per second, performance monitoring becomes more important if one also considers degradations due to photon–photon interactions, as discussed in previous chapters.

Performance monitoring at current bit rates is not as easy as it used to be. Instrumentation is complex, has large form factor, and is costly. As a result, it is not easily incorporated into every input unit of a communications system. Thus, performance monitoring relies on indirect means such as measuring BER from simple codes such as parity, to very sophisticated error detecting and correcting codes such as cyclic redundancy codes (CRC) and forward error correction (FEC). These codes have been added to each information frame to detect and correct and count errors and keep track of error seconds and other metrics defined in standards such as ANSI, IETF, ITU-T, and others. However, error detecting and correcting (EDC) codes are not solidly reliable because their detectability and correctability is limited (see Chapter 9). For example, a simple parity check can detect (and not correct) a single error but it fails to recognize two errors. A FEC code designed to detect sixteen and correct eight errors will not detect or correct more than these limits. More significantly, the FEC method requires several frames (or packets) to estimate the bit error rate and thus long monitoring periods that may compromise the overall system and network responsiveness to remedial actions (Figure 8.1). For example, if the bit rate of a channel is 10 Gb/s and the channel performance objective is 10^{-15} BER (meaning one error in 10^{15} bits), it may take more than 27 hours to measure.

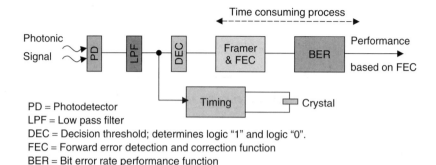

Figure 8.1. Current methods are based on FEC for estimating BER and performance of the incoming signal.

For 10^{-12} BER it may take 1.7 minutes, and for 10^{-10} it may take 1 second. However, even 1 second is a long period for networks that have hundreds or thousands of data streams, aggregate data rates at several terabits/s per fiber, and require switching to protection under in 50 ms. In addition, FEC codes add overhead to the signal (although this may be a small price to pay considering that it also corrects errors), and it also adds to the processing complexity of the input unit. Thus, a question arises: can we monitor the channel performance in real time and make adjustments in a proactive manner to avoid severe degradations and failures?

In this chapter, we address this question and provide a solution. We describe an estimation methodology that, based on statistical sampling, constructs "soft" histograms of the amplitude of incoming signal for "ones" and "zeros." These histograms represent probability distributions of the incoming symbols from which the bit error rate (BER), Q-factor, signal-to-noise ratio (SNR) and other performance parameters are estimated very fast (fraction of a second). In addition, based on this methodology a circuit is described that automatically estimates these parameters (BER, Q, SNR, etc.). The circuit lends itself to VLSI integration. Therefore, this method presents an affordable solution that can be deployable at the physical layer of each input unit of a communications system.

Estimation of bit error rate with eye diagrams is demonstrated with several exercises using the CD-ROM that accompanies this book (see Appendix B for a description of these exercises).

8.2 SAMPLING AND STATISTICAL HISTOGRAMS

Sampled data from experiments or observations are organized in groups of ranges and determine the frequency of occurrence of each group. This organization is used to construct the probability distribution of experimental data (Figure 8.2). From this data organization and distribution, the mean, standard deviation, and variance are derived for statistical data from the probabilistic data outlined in Chapter 7.

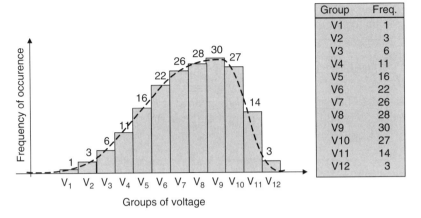

Figure 8.2. Raw data are organized in groups and the frequency of occurrence for each group is counted. This is used to construct the probability distribution.

For convenience, we repeat here the ones mostly used in this chapter:

- Root mean square: $\sqrt{\Sigma x_i^2/N}$, where x_i is a set of N numbers
- Mean value: $\mu = \Sigma x_j f(x_j)/\Sigma f(x_j) = \Sigma x_j f(x_j)/N$, where $f(x_j)$ is the frequency of occurrence (and not the probability of occurrence as in Chapter 7)
- Variance: $\text{var}(X) = \Sigma(x_i - \mu)^2 f(x_i)/N = \Sigma x_i^2 f(x_i)/N - \mu^2 = E(X^2) - \mu^2$, where $f(x_i)$ is the frequency of each group value x_i.
- Standard deviation: $\sigma = \sqrt{[1/(n-1)] \Sigma(x_i - x_{\text{mean}})^2} = \sqrt{\text{var}}$

Consider, for example, the data of Table 8.1 (as illustrated in Figure 8.2). From the data in Table 8.1, and for a total number of samples $N = 187$,

- Mean value: $\mu = \Sigma x_j f(x_j)/N = 1.5$ V
- Variance: $\text{var}(X) = \Sigma x_i^2 f(x_i)/N - \mu^2 = 2.515 - 2.250 = 0.265$ V^2
- Standard deviation: $\sigma = \sqrt{\text{var}} = 0.515$ V

8.3 STATISTICAL SAMPLING FOR BER

As bits (ones and zeros) arrive at the input, they arrive at different amplitudes due to temporal variations such as noise, attenuation, and other degradations discussed in Chapters 1 to 6. Consider sampling the bits at the optimum point in the bit period (typically within 0.5–0.6 T). Consider also that the frequency of occurrence of the sampled amplitudes is recorded. However, two thresholds may be defined, V2 and V1, so that those samples that are above a threshold V2 are recorded as "one" and those below V1 as "zero." The region between V2 and V1 represents the uncertain-

252   BER STATISTICAL MEASUREMENTS

ty of the decision. In the simplest case, V2 = V1. From the data obtained, the statistical parameters μ, σ^2, and σ are calculated as in the previous section.

8.4 ESTIMATING BER, Q-FACTOR, AND SNR

The samples of the incoming signal are grouped in memory locations. This may be viewed as a virtual histogram (Figure 8.4). From the stored and grouped data, the

Table 8.1. Calculations of raw data organized in groups

Group	Amplitude of group (x_j)	Freq. of occurence $f(x_j)$	$x_j f(x_j)$	x_j^2	$x_j^2 f(x_j)$
V1	0.2	1	0.2	0.04	0.04
V2	0.4	3	1.2	0.16	0.48
V3	0.6	6	3.6	0.36	2.16
V4	0.8	11	8.8	0.64	7.04
V5	1.0	16	16.0	1.0	16.0
V6	1.2	22	26.4	1.44	31.68
V7	1.4	26	36.4	1.96	50.96
V8	1.6	28	44.8	2.56	71.68
V9	1.8	30	54.0	3.24	97.2
V10	2.0	27	54.0	4.0	108
V11	2.2	14	30.8	4.84	67.76
V12	2.4	3	7.2	5.76	17.28
Sum		187	283.4		470.28

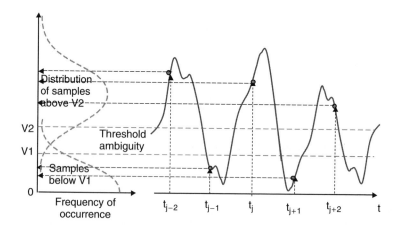

Figure 8.3. Periodic sampling of the incoming signal and tallying the frequency of occurrence of each sample. Distributions for "ones" and "zeros" are constructed from which the mean, variance, and standard deviation can be easily calculated.

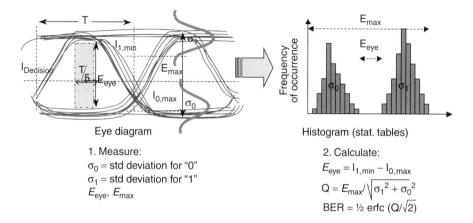

Figure 8.4. BER estimation based on eye diagram and statistical histograms.

statistical parameters μ_1, μ_0, σ_1, σ_0, $E_{1,\min}$, $E_{0,\max}$, E_{eye}, and $E_{\max}$ are obtained. These parameters are entered in the following relationships (see also Chapter 7):

$$E_{eye} = E_{1,\min} - E_{0,\max}$$

$$r = E_{\max}/E_{eye}$$

$$Q = (|\mu_1 - \mu_0|)/(|\sigma_1 - \sigma_0|)$$

$$BER = \tfrac{1}{2}\,\mathrm{erfc}\,(Q/\sqrt{2})$$

$$SNR_{appr} = (r+1)/(r-1)$$

$$R_{ext} = \mu_1/\mu_0$$

where R_{ext} is the extraction ratio. If the received signal contains overshoot and undershoot, $E_{\max}$ and E_{eye} are not exact. Then, an approximated $E^*_{\max}$ value may be used to compensate for it:

$$E^*_{\max} = E_{1,\mathrm{mean}} + k_1\sigma_1$$

where $k_1 = 1, 2,$ or 3 and $E_{1,\mathrm{mean}}$ is the mean value for logic "1." Similarly,

$$E^*_{eye} = (E_{1,\mathrm{mean}} - k_1\sigma_1) - (E_{0,\mathrm{mean}} + k_0\sigma_0)$$

where $k_0 = 1, 2,$ or 3, and $E_{0,\mathrm{mean}}$ is the mean value for logic 0.

In addition, from the estimated, V_p, and the calculated BER, the standard deviation of noise is calculated from

$$BER = \tfrac{1}{2}\{1 - \mathrm{erf}[V_p/(2\sigma_n\sqrt{2})]\}$$

Finally, from the measured input power, from the noise characteristics at the receiver, and the estimated SNR value, the noise figure is calculated:

$$NF = \frac{P_{si}/kT_0 \, df}{OSNR}$$

8.5 THE BER CIRCUIT

Based on the method of the previous section, a functional diagram of a circuit is described that is integrable with a single VLSI or with a VLSI and a microprocessor (the particular implementation is a designer's choice) (Figure 8.5).

With reference the functional block diagram of Figure 8.5, the incoming photonic signal is detected by a photodetector (PD) and is converted to an electrical signal. This is passed through a low-pass filter (LPF) to remove high-frequency noise. The electrical signal is routed to a sample and hold (S&H) function that periodically samples the signal and holds it for one period of the voltage value (V_s). The voltage, V_s, is sent to a voltage comparator that compares it with a reference voltage (V_{ref}). It is also sent to an analog-to-digital (A/D) converter function, where it is converted into digital codes.

The binary or digital codes from the A/D is sent to two functions, one that receives those that are greater than V_s, named "1s std dev," and another that receives those that are smaller than V_s, named "0s std dev"; this is controlled by the comparator that generates two enabling signals, $V_s > V_{ref}$ and $V_s < V_{ref}$, respectively. The "1s std dev" function builds statistical tables for those codes that correspond to $V_s > V_{ref}$ and from their distribution it determines the mean and the standard deviation for the "1s." Similarly, the "0s std dev" function builds statistical tables for those codes that correspond to $V_s < V_{ref}$ and from their distribution it determines the mean and the standard deviation for the "0s." These statistical parameters provide input to a functional block that calculates the Q factor and passes information signals to two

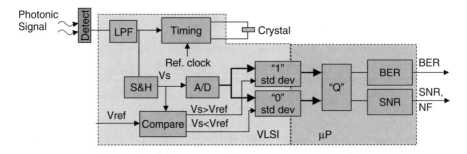

Figure 8.5. Functional block diagram of a circuit for the staistical estimation of bit error rate, Q-ratio, and signal-to-noise ratio. The actual implementation and circuit partitioning is a designer's choice.

functions labeled SNR and BER, which calculate the estimated signal-to-noise ratio and the bit error rate, respectively.

8.6 PERFORMANCE OF THE BER CIRCUIT

Because of the statistical nature of this method, a relatively small sample of periods (a few thousand), are required. For example, at 10 Gbit/s, 10,000 samples correspond to 1 μs; similarly, at 2.5 Gbit/s, 5,000 samples correspond to 2 μs. However, 1 or 2 μs is an extremely short time for circuitry to calculate the performance metric of the channel, BER, Q, and SNR. However, because errors are random, sampling may be statistical, such as one every 1000 sampled bits. Although this does not compromise the accuracy of the method, it does relax the circuitry switching requirements so that standard VLSI technology can be used. Thus, for the 10 Gbit/s case the channel performance parameters are estimated every millisecond, and for the 2.5 Gbit/s, every four milliseconds. Thus, a very fast performance evaluation is accomplished, which although not exact, provides a reasonable and very fast estimation of the channel performance, and does so on a continuous basis so that performance degradation can be detected as it happens. Moreover, the BER VLSI circuit may be included in each input of a system; this is a new approach that currently is not in use.

Thus, the BER estimation methodology, and the circuit based on it, provides several benefits for the quick statistical estimation of BER and SNR:

- It is based on statistical sampling of the incoming signal.
- It provides fast estimate of BER based on statistical sampling.
- It provides fast estimate of SNR and Q based on statistical sampling.
- It lends itself to implementation with integrated ICs.
- VLSI allows for per-channel, low-cost implementation.

8.7 OTHER PERFORMANCE METRICS

In addition to the BER, Q, and SNR metrics, the BER circuit may also include estimation functionality of other performance metrics such as:

- Errored Second (ES): a one-second interval that contains at least one error
- Severely Errored Second (SES): a one-second interval that contains >30% EBs
- Errored Block (EB): A block in which one or more bits are in error
- Burst Errored Seconds (BES): Any errored second that contains at least 100 errored EBs
- Background Block Error (BBE): An EB not occurring as part of an SES
- BBE Ratio: The ratio of EBs to total blocks during a fixed measurement inter-

val, excluding all blocks during SESs and unavailable time
- ES Ratio (ESR): The ratio of ESs to total seconds in available time during a fixed measurement interval
- SES Ratio: The ratio SESs to total seconds in available time during a fixed measurement interval

A block of bits is defined as $R\Delta\tau$ bits, where R is the bit rate and $\Delta\tau$ is the block period deined as $1/Bk$; Bk is the block rate or number of blocks per second.

Some objectives for design are:

- BER: 2×10^{-10}, excluding burst errors at the DS-1 to DS-3 interfaces
- Bursts: no more than 4 per day at DS-1 to DS-3 interfaces
- Long-term percentage of errored seconds: at the DS-1 and at the DS-3 interfaces, not to exceed 0.04%. This means 99.6% error-free seconds (EFS).

REFERENCES

1. M. Abramovitch and I. Stegun (Eds.), *Handbook of Mathematical Functions,* National Bureau of Standards, United States Department of Commerce, 1964.
2. T. Wildi, *Units and Conversion Charts,* 2nd ed., IEEE Press, 1995.
3. S.V. Kartalopoulos, *DWDM: Networks, Devices and Technology,* Wiley/IEEE-Press, 2003.
4. S.V. Kartalopoulos, *Fault Detectability DWDM: Toward Higher Signal Quality & System Reliability,* Wiley/IEEE-Press, 2001.
5. D. Derickson, *Fiber Optic Test and Measurement,* Prentice-Hall, 1998.
6. A.D. Whalen, *Detection of Signal in Noise,* Academic Press, 1971.
7. M. C. Jeruchim, P. Balaban, and K.S. Shanmugan, *Simulation of Communications Systems,* Plenum Press, 1992.
8. R. Marz, *Integrated Optics: Design and Modeling,* Artech House, 1995.
9. M.C. Jeruchim, "Techniques for estimating the bit error rate in the simulation of digital systems," *IEEE Journal on Selected Areas in Communications,* SAC-2, 1, 153–170, 1984.
10. K. Hinton and T. Stephens, "Modeling High-Speed Optical Transmission Systems," *IEEE Journal on Selected Areas in Communications,* 11, 4, 380–392, 1993.
11. A.J. Lowery, "Computer-Aided Photonics Design," *IEEE Spectrum,* 34, 4, 26–31, April 1997.
12. G. Bendelli, C. Cavazzoni, R. Giraldi, and R. Lano, "Optical Performance Monitoring Techniques," *Proceedings of the European Conference on Optical Communication, ECOC 2000,* 4, 113–116, 2000.
13. F. Abramovich and P. Bayvel, "Some Statistical Remarks on the Derivation of BER in Amplified Optical Communications Systems," *IEEE Trans. on Communications,* 45, 9, 1032–1034, 1997.
14. F. Matera and M. Settembre, "Role of Q-factor and of Time Jitter in the Performance

Evaluation of Optically Amplified Transmission Systems," *IEEE J. of Selected Topics in Quantum Electronics, 6,* 2, 308–316, 2000.

15. C. J. Anderson and J.A. Lyle, "Technique for Evaluating System Performance Using Q in Numerical Simulations Exhibiting Intersymbol Interference," *Electronics Letters, 30,* 1, 71–72, 1994.
16. N.S. Bergano, F.W. Kerfoot, and C.R. Davidson, "Margin Measurements in Optical Amplifier Systems," *IEEE Photonic Technical Letters, 5,* 3, 304–305, March 1993.
17. D. Marcuse, "Derivation of Analytical Expressions for the Bit-Error Probability in Lightwave Systems with Optical Amplifiers," *Journal Lightwave Technology, 8,* 1816–1823, 1990.
18. S. V. Kartalopoulos, "Factors Affecting the Signal Quality, and Eye-Diagram Estimation Method for BER and SNR in Optical Data Transmission," *Proceedings of the International Conference on Information Technology, ITCC-2004,* Las Vegas, April 5–7, pp. 615–619, 2004.
19. S. V. Kartalopoulos, "Per-port Circuit for Statistical Estimation of Bit Error Rate and Optical Signal to Noise Ratio in DWDM Telecommunications," *Proceedings of SPIE Fluctuation and Noise 2004 Conference,* Las Palmas, Maspalomas, Gram Canaria, Spain, May 25–28, 2004.
20. S.V. Kartalopoulos, "Apparatus and Method for Optical Pattern Detection," U.S. patent 6,617,566, issued 9/9/2003.
21. Z. Zalevsky and V. Eckhouse, "In-Channel OSNR and BER Measurement Using Temporal Superresolution Via Dynamic Range Conversion," *Journal of Lightwave Technology, 21,* 11, 2734–2738, November 2003.
22. H. Nishimoto et al., New Method of Analyzing Eye Patterns and its Application to High-Speed Optical Transmission System," *J. Lightwave Tech., 6,* 5, 678–685, 1988.
23. K.Mueller, N. Hanik, A. Gladisch, H-M Foisel, and C. Caspar, "Applications of Amplitude Histograms for Quality of Service Measurements of Optical Channel and Fault Identification," in *Proceedings of ECOC'98,* 707–708, September 1998.
24. N. Hanik et al., "Application of Amplitude Histograms to Monitor Performance of Optical Channels," *Electronic Letters, 5,* 403–404, 1999.
25. C.M. Weinert, "Histogram Method for Performance Monitoring of the Optical Channel," in *Proceedings of ECOC 2000, 4,* 121–11622, 2000.
26. I. Shake and H. Takara, "Averaged Q-Factor Method Using Amplitude Histogram Evaluation for Transparent Monitoring of Optical Signal-to-Noise Ratio Degradation in Optical Transmission System," *Journal Lightwave Technology, 29,* 8, 1367–1373, 2002.
27. I. Shake and H. Takara, "Transparent and Flexible Performance Monitoring Using Amplitude Histogram Method," in *OFC'02 Tech. Digest,* paper TuE1, pp. 19–21, OFC'02, Anaheim, CA, 2002.
28. I. Shake et al., "Determination of the Origin of BER Degradation Utilizing Asynchronous Amplitude Histograms," in *CLEO* Pacific Rim 2001, vol. II, 560–561, 2001.
29. I. Shake et al., "Bit Rate Flexible Quality Monitoring of 10 to 160 Gbit/S Optical Signals Based on Optical Sampling," *Electronic Letters, 25,* 2087–2088, 2000.
30. S. Kobayashi and Y. Fukuda, "A Burst-mode Packet Receiver with Bit-rate-discriminating Circuit for Multi-bit-rate Transmission System," in *IEEE LEOS'99,* WX4, 595–596, 1999.

STANDARDS

1. ANSI/IEEE 812-1984, *Definition of Terms Relating to Fiber Optics,* 1984.
2. Bellcore (currently Telcordia) GR-253-CORE, *Synchronous Optical Network (SONET) Transport Systems: Common Generic Criteria,* 1995.
3. CCITT Recommendation O.151, *Error Performance Measuring Equipment Operating at the Primary Rate and Above,* 1992.
4. CCITT Recommendation O.153, *Basic Parameters for the Measurement of Error Performance at Bit Rates Below the Primary Rate,* 1992.
5. ITU-T Recommendation G.650, *Definition and Test Methods for the Relevant Parameters of Single-Mode Fibres,* 1997.
6. ITU-T Recommendation G.661, *Definition and Test Methods for the Relevant Generic Parameters of Optical Fiber Amplifiers,* November 1996.
7. ITU-T Recommendation G.671, *Transmission Characteristics of Passive Optical Components,* 2001.
8. ITU-T Recommendation G.691, *Optical Interfaces for Single Channel STM-64, STM-256 and Other SDH Systems with Optical Amplifiers,* 10/2000.
9. ITU-T Recommendation G.692, *Optical Interfaces for Multichannel Systems with Optical Amplifiers,* 10/1998.
10. ITU-T Recommendation G.693, *Optical Interfaces for Intra-Office-Systems,* 11/2001.
11. ITU-T Recommendation G.826, *Error Performance Parameters and Objectives for International, Constant Bit Rate Digital Paths at or Above the Primary Rate,* 1993.
12. ITU-T Recommendation G.828, *Error Performance Parameters and Objectives for International, Constant Bit Rate Synchronous Digital Paths,* 3/2000.
13. ITU-T Recommendation G.829, *Error Performance Parameters Events for SDH Multiplex and Regenerator Sections,* 3/2000.
14. ITU-T G.957, *Optical Interfaces for Equipments and Systems Relating to the Synchronous Digital Hierarchy,* 1999.
15. ITU-T Recommendation O.150, *General Requirements for Instrumentation for Performance Measurements on Digital Transmission Equipment ,* 5/1996.

CHAPTER 9

Error Detection and Correction Codes

9.1 INTRODUCTION

Binary communications opened the road to many innovations in both implementation and theory. One of them was error detection and correction codes. But, what does this really mean to communications?

We answer this question with an example. Consider a channel that requires an expected bit error rate performance of 10^{-5}, or a single error in 100,000 bits received, at a bit rate of 2 Mbit/s and a maximum line length of 2 km. At this rate, 100,000 bits are received within 50 ms, and thus it is expected that there will be on average one error every 50 ms, or 20 errors per second.

Clearly, if we elongate the line length to four kilometers and keep the same rate, then because of attenuation and more noise added to the signal, the BER is expected to increase. For the sake of argument, let us assume that if we double the line length, then BER increases to 10^{-2}, or a single error every 100 bits received. Then, at 2 Mbit/s, there are expected to be on average one error every 50 µs, or 20,000 errors per second.

Now, assume that the signal includes an error detection and correction code such that on average it corrects one error every 200 bits received; that is, it corrects one error every 50 µs, or 20,000 errors per second. Such a code would have corrected all expected errors and made the 4 km line length errorless and possible to use. However, because the EDC adds approximately 10% of overhead to the original 2 Mbit/s signal, the line rate now becomes 2.2 Mbit/s, and this rate increase brings the BER to the original 10^{-5}. Similar calculations and arguments are also used for bit rates at 10 Gbit/s and 10^{-12} BER, when the fiber length needs to be increased from 60 km to 200 km.

As a consequence, efficient error correction "coding" allows for longer distances between transmitter and receiver, for lower transmitted power and weaker received signals, and for communication in noisy environments. Therefore, communications standards define the type of error detection and correction code to be used as well as the strategy that is best suited to the particular transmission applications.

The impact of FEC codes on reach for optical signals is simulated in one of the exercises using the CD-ROM that accompanies this book (see Appendix *B* for a description of these exercises).

9.2 CODE INTERLEAVING

As a signal is transmitted through a medium, some degrading sources (internal and/or external) may influence the signal in a predictable (attenuation, dispersion) or unpredictable (uncorrelated data) manner. Thus, as various signals are multiplexed to construct another at higher rate (such as SONET STS-N), noise errors may occur at any time and at any point over the length of the medium. As a tributary source may experience a burst of errors, it is necessary to spread these errors over time. One way to accomplish this is to interleave the bits of EDC codes from each tributary.

There are several methods of interleaving; for example, straight polling interleaving, bit-diagonal interleaving, block interleaving, interblock interleaving, convolutional interleaving, and others. Each method provides improvements over the others for the particular application they were intended for. A simple interleaving method polls in a sequential order several codes and interleaves them a bit at a time. Convolutional interleaving is also a polling method, but in this case the code symbols are shifted in a bank of registers, where each register provides M symbols more memory than the preceding one.

Interleaving is used extensively in wireless, wired, and optical transmission. For example, error protection in SONET is provided by the overhead bytes B1 (section BIP-8), B2 (line BIP-8), and B3 (path BIP-8), which use interleaving; BIP stands for byte interleaved parity. Thus, per the SONET GR-253 standard:

- B1 uses even parity and it is calculated over all bits of the previous STS-N frame after scrambling and placed in the B1 byte of the current STS-N frame before scrambling
- The N line B2 codes in a STS-N are intended to form a single error-monitoring facility capable of measuring BER up to 10^{-3}, independent of the value of N.
- B3 uses even parity. BIP-8 is calculated over all bits of the previous STS SPE (783 bytes for an STS-1 SPE or NX783 for an STS-Nc SPE, regardless of pointer adjustments) before scrambling, and placed in the B3 byte of the current STS SPE before scrambling.

9.3 ERROR DETECTION AND CORRECTION STRATEGIES

Error detection and correction codes (EDC) can be as simple as parity or as complex as based on a complex theory of binary generating polynomials. In general, EDCs can be divided into two large categories: Hamming and convolutional codes.

9.3.1 Parity

An example of a parity code (also known as a *redundancy code*) is the parity bit in every ASCII character. The latter is an eight-bit code that consists of seven data bits and one (the eighth) parity bit. Parity is defined as odd or as even. If odd parity is specified, then the parity bit is "one" if the number of "ones" in the seven bits is odd; if even parity is specified, then the parity bit is "one" if the number of "ones" in the seven bits is even. The parity bit performs a simple check that is able to detect a single error or an odd number of errors, but it does not detect two or an even number of errors and it does not correct errors. A parallel parity circuit is constructed with modulo-2 or exclusive-or gates (Figure 9.1).

Another parity check example is if we consider a block of characters (Figure 9.2). Thus, if the block is transmitted in the sequence as

$$\{code1\}\{code2\}\{code3\}\{code4\}\{code5\}\{code6\}\{code7\}\{parity\ code\}$$

where each code includes its own parity bit, then at the receiving end certain errors can be detected and also corrected (Figures 9.3 and Figure 9.4).

However, the reliability of this EDC strategy collapses as errors appear in specific locations (Figure 9.5). As a consequence, although parity adds value, neverthe-

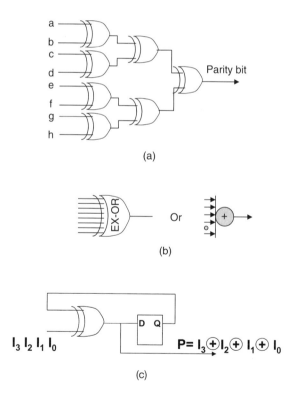

Figure 9.1. Two-input exclusive-or gates construct a parallel parity generator (a). The parity generator model is shown as an eight-input exclusive-or (b). A minimal serial parity generator employs an exclusive-or gate and a RS- or D-type flip-flop (c).

262 ERROR DETECTION AND CORRECTION CODES

Code	7-bit code	Parity
Code 1	0001101	1
Code 2	0111001	0
Code 3	0011101	0
Code 4	0111100	0
Code 5	0011010	1
Code 6	0110101	0
Code 7	0101010	1
Parity code:	0010100	1

Figure 9.2. A block of codes with horizontal and vertical parity check.

Code	7-bit code	Parity
Code 1	0001101	1
Code 2	0111001	·0
Code 3	0011101	0
Code 4	011**0**100	**0** ⇐
Code 5	0011010	1
Code 6	0110101	0
Code 7	0101010	1
Parity code:	001**0**100	1

Figure 9.3. The received block of codes with an errored bit (underlined). The corresponding horizontal and vertical bits locate the error, which can then be corrected.

Code	7-bit code	Parity
Code 1	0001101	1
Code 2	01110**11**	**0** ⇐
Code 3	0011101	0
Code 4	011**0**100	**0** ⇐
Code 5	0**1**11010	**1** ⇐
Code 6	0110101	0
Code 7	0101010	1
Parity code:	00**1****0**100	1

Figure 9.4. The received block of codes with three errored bits (underlined). The corresponding horizontal and vertical bits locate the error, which can then be corrected.

9.3 ERROR DETECTION AND CORRECTION STRATEGIES

Code	7-bit code	Parity
Code :	0001101	1
Code 2	0111001	0
Code 3	0011101	0
Code 4	0110100	0 ⇐
Code 5	0111000	1 ⇐ ?
Code 6	0111101	0 ⇐
Code 7	0101010	1
Parity code:	0010100	1

Figure 9.5. A case of multiple errors that cannot be reliably detected or corrected.

less it represents 12.5% of bandwidth overhead since for each seven bits one bit is added. It also adds more overhead if the parity code is also added; the amount of this overhead depends over how many code words the parity (or redundant code) is calculated (7 in the above example).

9.3.2 Hamming Codes

The circuit for the serial calculation of parity in Figure 9.1c is extended to more complex "shift-register" structures; such structures are used in serial error-detection/correction code implementation. Figure 9.6 illustrates a simple three-stage shift-register with an exclusive-or in the feedback path.

Hamming codes belong to the category already outlined. They is based on Modulo-2 (exclusive-OR) to calculate a redundant code that is able to detect and correct errors. Hamming codes are denoted as $H(n, k)$, where n is the sum of information and redundant bits, and k the protected information bits in a code.

To summarize the theoretical derivation of a Hamming code, consider an eight-bit byte that is comnposed of four information bits ($I0, I1, I2$, and $I3$), three redundant bits (H_0, H_1, and H_2) and a (even) parity bit P. In this case, the protected information length is 4, the Hamming code length is 7, and this is a $H(7, 4)$. Notice that the Hamming distance* is $d = 3$.

Based on the theory of Hamming cyclic linear block codes, the Hamming redundant bits in vectorial notation **H** [H0, H1, H2] is evaluated from the product of the information vector **I** [I0, I1, I2, I3] and the (4 × 4) generating matrix **G**:

$$\mathbf{H} = \mathbf{I} \times \mathbf{G}$$

Now, the derivation of the generating matrix is beyond the purpose of this book,

---

*Hamming distance between two binary code words of same length is the number of bits in which the two codewords differ. For example, the actual distance between 110 it and 111 is 1 as they differ by one bit, and between 101 and 110 it is two. However, the maximum distance between these two three-bit code words is 3 since the maximum possible distance between 000 and 111 is 3.

264 ERROR DETECTION AND CORRECTION CODES

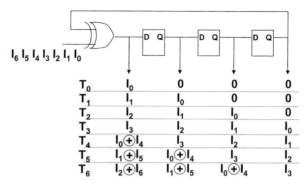

Figure 9.6. The parity calculation of serial streams in communications, Figure 9.1c, becomes the starting point for in-line error-detection/correction codes. As an exercise, verify the output of each element in this figure for seven consecutive periods.

but for completeness we illustrate the procedure with an actual step-by-step example.

Example
Consider the Hamming code $H(11, 7)$ able to correct a single error with four redundant bits. Also consider the seven-bit information code 1010010.

Step 1
Assuming transmission from left to right, this is written as an 11-bit code word:

$$\begin{array}{ccccccccccc} 11 & 10 & 9 & 8 & 7 & 6 & 5 & 4 & 3 & 2 & 1 \\ 1 & 0 & 1 & x & 0 & 0 & 1 & x & 0 & x & x \end{array}$$

where locations ($2^3 = 8$, $2^2 = 4$, $2^1 = 2$, and $2^0 = 1$) marked with x will host the redundant bits.

Step 2
The redundant bits are calculated according to the following alrgorithm. Convert in hex code those locations (from step 1) that have a one bit, and calculate the parity vertically:

$$\begin{array}{rcl} 11 & \Rightarrow & 1011 \\ 9 & \Rightarrow & 1001 \\ 5 & \Rightarrow & 0101 \\ \hline \text{parity} & & 0111 \end{array}$$

9.3 ERROR DETECTION AND CORRECTION STRATEGIES

Step 3
Insert the parity bits back into the locations with *x* (of step 1) in the same order (left to right); they are underlined for easy identification:

```
11  10  9  8  7  6  5  4  3  2  1
 1   0  1  0  0  0  1  1  0  1  1
```

When the 11-bit code is received, calculate the parity as in step 2, but including all bits in the codeword with ones. If no error has occurred during transmission, the parity code should be zero.

Step 4A

$$
\begin{aligned}
11 &\Rightarrow 1011\\
9 &\Rightarrow 1001\\
5 &\Rightarrow 0101\\
4 &\Rightarrow 0100\\
2 &\Rightarrow 0010\\
1 &\Rightarrow 0001\\
\hline
\text{parity} &\quad 0000
\end{aligned}
$$

A zero parity indicates no errored bits.

Step 5A
Remove redundant bits. Now assume that there has been a bit error in location 9 (shown in parentheses):

```
11  10   9  8  7  6  5  4  3  2  1
 1   0  (0) 0  0  0  1  1  0  1  1
```

Step 4B

$$
\begin{aligned}
11 &\Rightarrow 1011\\
9 &\Rightarrow 0000\\
5 &\Rightarrow 0101\\
4 &\Rightarrow 0100\\
2 &\Rightarrow 0010\\
1 &\Rightarrow 0001\\
\hline
\text{parity} &\quad 1001
\end{aligned}
$$

Clearly, the parity now is not zero, indicating that an error has occurred and also the parity point to the location where the error has occurred: $1001 => 9$.

266 ERROR DETECTION AND CORRECTION CODES

Step 5B
Correct the errored bit in location 9 and remove the redundant bits.

9.3.3 Convolutional Codes

Convolutional codes are generated by shifting the bit stream through a register. The shift register is organized in K groups, each group having k bits length. Specific outputs from each of the K registers (x^i) are exclusive-ored to generate n parity bits that construct an n-bit CC(n, k, K) convolutional code (Figure 9.7). Each of the n bits of the n-bit code is generated by a polynomial g_i. Therefore, there are n generating polynomials, $g_1, g_2, \ldots, g_n$.

An example of a convolutional code with $r = 2$, $k = 1$, and $K = 3$ or CC(2, 1, 3) and the two generating polynomials are shown in Figure 9.8.

In practice,, the incoming signal is "scrambled." However, if the received signal is passed through the same convolutional process with the same generating polynomials, then a descrambling process takes place and the original signal is restored. An observation (which follows from the example in Figure 9.7) is that k symbols are encoded to n symbols.

9.3.4 Block Codes

Block codes consider a block of k information symbols uniquely encoded into a block of n symbols. The ratio k/n determines the redundancy of the code and is known as the coding rate of the code. The late 1950s saw a feverish activity to find

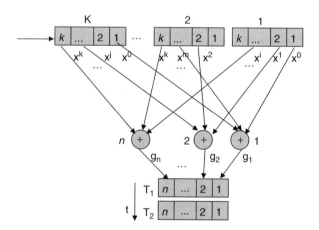

Figure 9.7. Convolutional CC(n, k, K) codes. K registers, each k-bits long, and n parity generators constuct an n-bit convolutional code word according to an algorithm. As bits shift across the register (from left to right), and for every period, a different code word is generated.

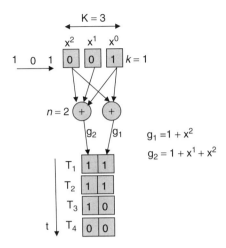

Figure 9.8. Example of CC(2, 1, 3) convolutional codes. Logic diagram (left) and generating polynomials (right).

the best error correcting codes. It is of interest that the correcting capability of block coding was introduced by Hamming (1950). Hocquenghem (1959) discovered a subclass of multiple error correcting codes as did Bose and Ray-Chaudhuri (1960); this was named the Bose–Chaudhuri–Hocquenghem (BCH) code. Peterson (1960) discovered the cyclic structure of these codes, and Reed and Solomon (1960) developed encoders that yielded the maximum separable distance between code words and improved efficiency, named RS encoders. This research activity continued and still continues for both optical and wireless transmission.

The study of block codes starts with the notion of finite fields (that is, fields with a limited number of symbols) that describe finite sets of q (0 to $q - 1$) elements and two operations, addition and multiplication; these are known as *Galois fields* (GF, after the French mathematician Evariste Galois who developed them at the age of 20 in 1832) and are denoted as $GF(q)$. As an example, Table 9.1 provide the addition and multiplication operations for $GF(5)$.

Notice that the result of multiplication or addition in $GF(q)$ cannot exceed the

Table 9.1. Multiplication and addition for $GF(5)$

×	0	1	2	3	4	+	0	1	2	3	4
0	0	0	0	0	0	0	0	1	2	3	4
1	0	1	2	3	4	1	1	2	3	4	0
2	0	2	4	1	2	2	2	3	4	0	1
3	0	3	1	4	2	3	3	4	0	1	2
4	0	4	3	2	1	4	4	0	1	2	3

268 ERROR DETECTION AND CORRECTION CODES

number $q - 1$. For example, $2 \times 4 = 3$ and $2 + 4 = 1$. The product is produced by mentally multiplying $2 \times 4 = 8$ and then subtracting 5 (or multiples of it) to produce 3. The sum is produced by mentally adding $2 + 4 = 6$ and then subtracting 5 (or multiples of it) to produce 1. This is known as *modulo-5* operation. When $q = 2$, then we have the well-known modulo-2 or exclusive-or logic function shown in Table 9.2.

Based on this, a $GF(12)$ is shown in Table 9.3. The purpose of this is to demonstrate that as q increases, then certain repetitive patterns start forming, which when used appropriately, can be used to develop encoding procedures useful in communications.

From Table 9.3, we make three observations (more observations may be made but we leave them for the enjoyment of the reader):

1. In the multiplication table, the (underlined) row #3 has a repetitive pattern, whereas the (underlined) row #7 is random.
2. Moreover, corresponding rows and columns have the same patterns of numbers.
3. In the addition table, the pattern of numbers is diagonally repetitive.

Now, some observations lead to the following:

Table 9.2. The Modulo-2 or exclusive-or is a $GF(2)$

	0	1
0	0	1
1	1	1

Table 9.3. Multiplication and addition for $GF(12)$

×	0	1	2	3	4	5	6	7	8	9	10	11	+	0	1	2	3	4	5	6	7	8	9	10	11
0	0	0	0	0	0	0	0	0	0	0	0	0	0	0	1	2	3	4	5	6	7	8	9	10	11
1	0	1	2	3	4	5	6	7	8	9	10	11	1	1	2	3	4	5	6	7	8	9	10	11	0
2	0	2	4	6	8	10	0	2	4	6	8	10	2	2	3	4	5	6	7	8	9	10	11	0	1
3	**0**	**3**	**6**	**9**	**0**	**3**	**6**	**9**	**0**	**3**	**6**	**9**	3	3	4	5	6	7	8	9	10	11	0	1	2
4	0	4	8	0	4	8	0	4	8	0	4	8	4	4	5	6	7	8	9	10	11	0	1	2	3
5	0	5	10	3	8	1	6	11	4	9	2	7	5	5	6	7	8	9	10	11	0	1	2	3	4
6	0	6	0	6	0	6	0	6	0	6	0	6	6	6	7	8	9	10	11	0	1	2	3	4	5
7	**0**	**7**	**2**	**9**	**4**	**11**	**6**	**1**	**8**	**3**	**4**	**1**	7	7	8	9	10	11	0	1	2	3	4	5	6
8	0	8	4	0	8	4	0	8	4	6	2	4	8	8	9	10	11	0	1	2	3	4	5	6	7
9	0	9	6	3	0	9	6	3	0	9	6	3	9	9	10	11	0	1	2	3	4	5	6	7	8
10	0	10	8	6	4	2	0	4	2	6	4	2	10	10	11	0	1	2	3	4	5	6	7	8	9
11	0	11	10	9	8	7	6	1	4	3	2	1	11	11	0	1	2	3	4	5	6	7	8	9	10

- If each row represents a vector, then there is a logical dependency from vector to vector that depends on the $GF(q)$ definition and rules.
- There is an element α in $GF(q)$, such that Modulo-$q\{\alpha^m\}$, $m = 1$ to $q - 1$, can represent every element (except zero) in the field. This is known as a *primitive element* and in the case of $GF(5)$ there are two: $\alpha = 2$ since $2^1 = 2$, $2^2 = 4$, $2^3 = 3$, and $2^4 = 1$, and $\alpha = 3$ since $3^1 = 3$, $3^2 = 4$, $3^3 = 2$, and $3^4 = 1$.
- The primitive element α gives rise to primitive polynomials for the $GF(q)$ fields, which are in the form $p(z) = \Sigma z^m$, such that α^m is the Modulo-$q\{p(z)\}$; the description of the generation of these polynomials is beyond the purpose of this chapter (the interested reader is encouraged to read one of the books listed in the references section).

Having described the Galois fields and their interesting properties, one could develop more complex fields by changing certain rules concerning addition and multiplication and study the patterning effects. Although this may seem to be just a stimulating mental exercise, neerthless if the efficiency of the generated patterns to real-life problems is proven to be applicable, then one has discovered a new coding process.

9.3.5 Cyclic Codes

For an (n, k) linear code C, k information symbols are encoded into an n-symbol code word. Now, assume that this is accomplished using shift registers. Then, this is a *cyclic code* if every cyclic shift of a vector in C yields a code vector also in C. That is, if we have an n-tuple vector $\mathbf{V} = (\mathbf{v}_{n-1}, \ldots, \mathbf{v}_1, \mathbf{v}_0)$ and it is cyclically shifted to the left by one position, another n-tuple vector $\mathbf{V}^{(1)}$ is obtained, $\mathbf{V}^{(1)} = (\mathbf{v}_{n-2}, \ldots, \mathbf{v}_1, \mathbf{v}_0, \mathbf{v}_{n-1})$, and so on.

The cyclic codes have some interesting properties that make them applicable to error detection and correction:

- For an (n, k) linear code C, there are 2^k code word polynomials $c(z)$. The codeword polynomials of degree $(n - 1)$ are encoded by a generating polynomial $g(z)$ of degree $(n - k)$.
- The generating polynomial $g(z)$ of an (n, k) code is a factor of $(z^n + 1)$.
- For any cyclic code with a generating polynomial $g(z)$ such that $z^n + 1 = g(z)h(z)$, $h(z)$ is the parity-check polynomial.

Thus, if the information sequence is represented by a set of polynomials $i(z)$ of degree $(k - 1)$, then the information polynomials are mapped into a set of code word polynomials $c(z)$ using the generator polynomial $g(z)$ in a *nonsystematic encoding* process described by

$$c(z) = i(z)g(z)$$

Notice that the code word does not contain a copy of $i(z)$, and, hence, is nonsystematic. Nonsystematic implies that the information word needs to be processed (or disturbed) by the decoder at the receiver in order to calculate the parity symbols.

Alternatively, if $i(z)$ is inserted in the high-order coefficients of the code word $c(z)$ and the parities (or redundant symbols) are appended to the low-order coefficients (so that the information word is not disturbed in any way in the encoder), then there is *systematic encoding* described by

$$c(z) = i(z)z^{n-k} + b(z)$$

9.3.6 Binary BCH Codes

The Bose–Chaudhuri–Hocquenghem (BCH) codes are a class of cyclic block codes with multiple error detection and correction capability. When they are defined in a $GF(2)$ field, then binary BCH codes are applicable to digital communications.

In general, a BCH code accepts k information symbols and produces an n-symbol code word. If the code word corrects t random errors, the code is called t-error-correcting and it is denoted as $BCH(n, k, t)$. In this case, the distance between code words is $d = 2t + 1$. Because BCH codes are cyclic, they can be constructed by their generating polynomials.

The conclusion of such codes is that for k information bits, $n - k$ additional bits are generated to produce an n-symbols-long code word to correct t bits. Thus, a BCH(15, 7, 2) means that seven bits of information will become a 15-bit code word (by adding 8 bits to 7) in order to be able to correct any two bits in the code word. This code adds a 114.28% overhead increase to the original code. Similarly, the BCH(15, 5, 3) code is able to correct three errors, but in this case the original 5 information bits became 15 code word bits; that is, an overhead of 300% is added. A consequence, the line data rate increases by the amount of the added overhead.

The overhead added to a code is a metric of the efficiency of the particular code. Thus, although BCH encoding set the mindset for error detection and correction using cyclic codes, more efficient codes were needed.

9.3.7 Reed–Solomon Codes and FEC

An improved version of cyclic encoders with maximum separable distance and improved efficiency are the Reed–Solomon $\{RS(n/k)\}$ codes, where k is the number of information bytes and $n - k$ is the error correction code (in bytes); these codes belong to the subclass of systematic linear cyclic block codes, and because the Reed–Solomon RS(255,239) FEC code operates on 8-bit symbols, it is a nonbinary code. However, the coding structure is compatible with binary transmission.

The $RS(n/k)$ error detection and correction code with $n = 255$ bytes and $k = 239$ bytes is the code adopted by ITU-T for the Optical Transport Network standard (G.

9.3 ERROR DETECTION AND CORRECTION STRATEGIES

709). A RS(255, 239) coder calculates 16 check bytes over 239 byte symbols, which append to the 239 bytes (Figure 9.9). The appended 16 check bytes are known as the *forward error correction code* (FEC). Thus, the overhead added by it is only 16/239, or approximately 7%.

The generator polynomial of the A RS(255, 239) code is

$$G(z) = \prod_{i=0}^{15}(z - \alpha^i)$$

where α is a root of the binary primitive polynomial $x^8 + x^4 + x^3 + x^2 + 1$.

A data byte $(d_7, d_6, \ldots, d_1, d_0)$ is identified with the element $d_7 \cdot \alpha^7 + d_6 \cdot \alpha^6 + \ldots + d_1 \cdot \alpha^1 + d_0$ in $GF(256)$, the finite field with 256 elements.

In OTN, FEC encoding is performed in the terminal transmission equipment (TTE) at each STM-16 signal and before interleaving M STS-16 tributaries. Thus, the FEC encoder accepts information bits from which it computes redundant symbols, which it appends to the information bits, producing encoded data at a higher bit rate.

Decoding is performed after deinterleaving the signal. The FEC decoder performs the error detection and error correction while extracting the redundancy to regenerate the data that was encoded by the FEC encoder. The BER at the input is expected to be in the range from 10^{-3} to 10^{-15}.

ITU-T G.709 and G.975 provide the theoretical correcting performance relationship for the RS(255, 239) code, taking into account the line BER (BER_{Input}) before FEC function correction and the line BER (BER_{Output}) after FEC function correction, as follows:

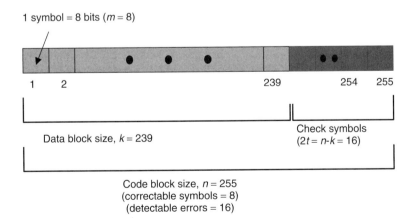

Figure 9.9. The next-generation optical transport network ITU-T standards defines a Reed-Solomon code RS(255, 239) that is appended to the information field; it is known as the forward error correction (FEC) code (G.709 and G.975).

$$\begin{cases} P_{UE} = \sum_{i=9}^{N} \frac{i}{N} \cdot \binom{N}{i} \cdot P_{SE}^{i} \cdot (1 - P_{SE})^{N-i} & \text{with } N = 255 \\ BER_{\text{Input}} = 1 - (1 - P_{SE})^{1/8} \\ BER_{\text{Output}} = 1 - (1 - P_{UE})^{1/8} \end{cases}$$

where, P_{UE} is the probability of an uncorrectable error, P_{SE} is the probability of a symbol (byte) error, and N is the code word length (255).

To complete the above set of relationships, we also state the BER in terms of the Q-factor:

$$BER_{\text{Output}} = \tfrac{1}{2}\,\text{erfc}\,(Q/\sqrt{2})$$

Based on the before and after FEC, a BER improvement or gain may be calculated that is indicative of the performance of a particular EDC code (from which Q may also be calculated). Figure 9.10 (adapted from ITU-T G.975) and Table 9.4 show the performance of the RS(258, 239) code before and after FEC correction.

9.4 SUMMARY

In this chapter, we have just scratched the surface of a very important issue in transmission. We have briefly outlined the ideas of codes that are capable of detecting and correcting code words and we explained how EDCs, by correcting errors, maintain the expected channel performance. Besides the codes mentioned here, there are

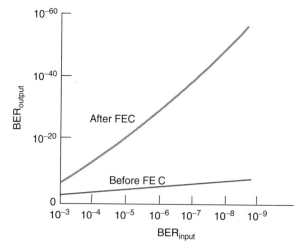

Figure 9.10. The error correction capability of FAC RS(255, 239) is dramatic. (Adapted from ITU-T standard G.975.)

Table 9.4. BER at the input and output of the FEC decoder (at 2.5 Gbit/s)

BER_{Before}	BER_{After}
10^{-4}	5×10^{-15}
10^{-5}	6.3×10^{-24}
10^{-6}	6.4×10^{-33}

other powerful EDC codes, such as Trellis codes, enhanced-FEC, 3-D, turbo codes, and others that are being aggressively researched. The interested student of EDCs can find many recent papers in this area.

In optical communications, FEC codes, in addition to detecting and correcting errors, also report errors so that as degradations become evident, maintenance may be undertaken for before a hard failure occurs. The BER estimation method described in Chapter 8, although it has no error correction capability, provides a much faster qualitative indication of the signal quality and channel performance with no overhead on the transmission rate. Therefore, when the BER circuit works in concert with the FEC decoding, a more powerful method is constructed that both detects early degradations and corrects errors.

REFERENCES

1. J. Adamek, *Foundations of Coding*, Wiley, 1991.
2. J.L. Massey, *Threshold Decoding*, MIT Press, 1963.
3. G.D. Forney, *Concatenated Codes*, MIT Press, 1966.
4. E.R. Berlekamp, *Algebraic Coding Theory*, McGraw-Hill, 1968; revised edition, Aegean Park Press, 1984.
5. W.W. Peterson and E.J. Weldon, *Error Correcting Codes*, MIT Press, 1972.
6. F.J. MacWilliams and J.A. Sloane, *The Theory of Error-Correcting Codes*, North-Holland, 1977.
7. N. Glover and T. Dudley, *Practical Error Correction Design for Engineers*, 2nd ed., Data Systems Technology Corp., Bloomfield, CO, 1998.
8. V. Pless, *Introduction to the Theory of Error-Correcting Codes*, 3rd ed., Wiley, 1998.
9. R.E. Blahut, *Theory and Practice of Error Control Codes*, Addison-Wesley, 1983.
10. S. Lin and D.J. Costello, Jr., *Error Control Coding: Fundamentals and Applications*, Prentice Hall, 1983.
11. R.W. Hamming, "Error Detecting and Error Correcting Codes," *Bell System Technical Journal, 28,* pp. 147–160, 1950.
12. P. Elias, "Coding for Noisy Channels," in *IRE Conv. Rec.* part 4, pp. 37–47, 1955.
13. Hocquenghem, "Codes Correcteurs d'Erreurs," vol. 2, Chiffres, (pp. 147–156), September 1959.
14. R.C. Bose and D.K. Ray-Chaudhuri, "On a Class of Error Correcting Binary Group Codes," *Information and Control, 3,* 68–79, March 1960.

15. R.C. Bose and D.K. Ray-Chaudhuri, "Further Results on Error Correcting Binary Group Codes," *Information and Control, 3,* 279–290, September 1960.
16. I.S. Reed and G. Solomon, "Polynomial Codes over Certain Finite Fields," *J. Soc. Ind. Appl. Math., 8,* 300–304, June 1960.
17. W.W Peterson, "Encoding and Error Correction Procedures for the Bose-Chaudhuri Codes," *IRE Trans. Inform. Theory, IT-6,* 459–470, September 1960.
18. D. Gorenstein and N. Ziegler, "A Class of Cyclic Linear Error-Correcting Codes in p^m Symbols," *J. Soc. Ind. Appl. Math., 9,* 107–214, June 1961.
19. R.M. Fano, "A Heuristic Discussion of Probability Coding," *IEEE Trans. Information. Theory, IT-9,* 64–74, April 1963.
20. J.L. Masey, "Step-by-Step Decoding of the Bose-Chaudhuri-Hocquenghem Codes," *IEEE Trans. Info. Theory, IT-11,* 580–585, 1965.
21. A.J. Viterbi, "Error Bounds for Convolutional Codes and an Asymptotically Optimum Decoding Algorithm," *IEEE Trans. Info. Theory, IT-13,* 260–269, April 1967.
22. J.L. Masey, "Shift-Register Synthesis and BCH Decoding," *IEEE Trans. Info. Theory, IT-15,* 122–127, January 1969.
23. A.J. Viterbi, "Convolutional Codes and Their Performance in Communications," *IEEE Trans. Commun. Technol., COM-19,* 5, 751–772, October 1971.
24. G.D. Forney, "Burst-Correctign Codes for the Classic Burst Channel," *IEEE Trans. Comm. Technol., COM-19,* 5, 772–781, October 1971.
25. E. Prange, "Cyclic Error-Correcting Codes inn Two Symbols," *AFCRC-TN-57, 103,* Air Force Cambridge Research Center, Cambridge, Mass, 1972.
26. G.D. Forney, "The Viterbi Algorithm," *Proceedings of the IEEE, 61,* 3, 268–278, March 1973.
27. E.R. Berlekamp, "The Technology of Error-Correcting Codes," *Proceedings of the IEEE, 68,* 5, 564–592, May 1980.
28. Srivastava, S. Vandris, A. Velingker, and S.R. Goldman, "Enhanced Forward Error Correction Algorithms," in *Technical Proceedings of NFOEC,* pp. 653–662, 2002.

STANDARDS

1. ANSI/IEEE 812-1984, *Definition of Terms Relating to Fiber Optics,* 1984.
2. ANSI T1X1.5/99-002, *A Proposal for Providing Channel-Associated Optical Channel Overhead in the OTN,* Lucent Technologies (January 1999), http://www.t1.org/index/0816.htm.
3. ANSI T1X1.5/99-003, *A Proposal Implementation for a Digital "Wrapper" for OCh Overhead,* Lucent Technologies (January 1999), http://www.t1.org/index/0816.htm.
4. ANSI T1X1.5/99-004, *Optical Channel Overhead Carried on the Optical Supervisory Channel,* Lucent Technologies (January 1999), http://www.t1.org/index/0816.htm.
5. IEC Publication 1280-2-1, *Fibre Optic Communication Subsystem Basic Test Procedures; Part 2: Test Procedures for Digital Systems; Section 1—Receiver Sensitivity and Overload Measurement.*
6. IEC Publication 1280-2-2, *Fibre Optic Communication Subsystem Basic Test Procedures; Part 2: Test Procedures for Digital Systems; Section 2—Optical Eye Pattern, Waveform and Extinction Ratio Measurement.*

7. IEEE 802.3ab, *1000BaseT*.
8. Internet study group: *http://www.internet2.edu*.
9. ITU-T Recommendation G.650, *Definition and Test Methods for the Relevant Parameters of Single-Mode Fibres*, 1996.
10. ITU-T Recommendation G.702, *Digital Hierarchy Bit Rates*, 1988.
11. ITU-T Draft Rec. G.798, *Characteristics of Optical Transport Networks (OTN) Equipment Functional Blocks*, October 1998.
12. ITU-T Rec. G.805, *Generic Functional Architecture of Transport Networks*, October 1998.
13. ITU-T G.825, *The Control and Wander Within Digital Networks which are Based on the Synchronous Digital Hierarchy (SDH)*, March 2000.
14. ITU-T Rec. G.873, *Optical Transport Network Requirements*, October 1998.
15. ITU-T Recommendation G.957, *Optical Interfaces for Equipments and Systems Relating to the Synchronous Digital Hierarchy*, June 1995.
16. ITU-T G.975, *Forward Error Correction for Submarine Systems*, November 1996.
17. Telcordia (formerly Bellcore) GR-253-CORE, *Synchronous Optical Network (SONET) Transport Systems: Common Generic Criteria*, Issue 2, December 1995.

APPENDIX A

Statistical Noise in Communications

Communications theory combines signal processing, statistics, and probabilities to derive certain powerful relationships of serious concern in communications. In this appendix we provide a rigorous mathematical analysis of the probabilistic estimation of noise. The one-sided spectral density is expressed by

$$2G_n(f) = K|H(\omega)|^2$$

The narrow-band noise is expressed by

$n(t) = \Sigma \sqrt{[4G_n(f)\,\Delta f]} \cos(\omega_i t + \theta_i)$, where the sum is from $i = 1$ to infinity

To find the narrowband characteristics of noise, let $\omega_1 = (\omega_1 - \omega_0) + \omega_0$, then,

$$n(t) = \{\Sigma \sqrt{[4G_n(f)\,\Delta f]} \cos[(\omega_1 - \omega_0)t + \theta_1)]\} \cos \omega_0 t$$
$$- \{\Sigma \sqrt{[4G_n(f)\,\Delta f]} \sin[(\omega_1 - \omega_0)t + \theta_1)]\} \sin \omega_0 t$$
$$= x(t)\cos \omega_0 t - y(t)\sin \omega_0 t$$

It turns out that the mean square values are

$$\langle x^2 \rangle = \langle y^2 \rangle = N = \langle n^2 \rangle$$

where x and y are independent variables, or

$$\langle x^2 \rangle = \Sigma[2G_n(f)\,\Delta f]$$

where the sum is over 1, Δf at the limit is replaced by df, and T tends to infinity, so that

$$\langle x^2 \rangle = N$$

and

$$\langle y^2 \rangle = N$$

Thus,

$$n(t) = x(t)\cos \omega_0 t - y(t)\sin \omega_0 t = r(t)\cos(\omega_0 + \theta)t$$

where $r = \sqrt{x^2 + y^2}$ is the distribution envelope, and $\theta = \tan^{-1}(y/x)$ is the phase. The distribution of the envelope is found as follows:

$$p(x, y) = [\exp(-r^2/2N)]/2\pi N$$

or, expressed in polar coordinates,

$$p(r, \theta)dr\, d\theta = \{[r \exp(-r^2/2N)]/2\pi N\} dr\, d\theta$$

Averaging over all θ and integrating from 0 to 2π, then,

$$P(r) = \text{Int}\{p(r, \theta)d\theta\} = [r \exp(-r^2/2N)]/N$$

This is known as the Rayleigh distribution (Figure A.1) and it represents the probability distribution of the envelope of noise alone in the narrowband bandpass filter.

The Rayleigh distribution is normalized by integrating from 0 to infinity over r and set to unity:

$$\text{Int}[P(r)dr] = \text{Int}[r \exp(-r^2/2N)]dr/N = 1$$

Now, the resultant signal plus noise is

$$U(t) = n(t) + A \cos \omega_0 t = (x + A) \cos \omega_0 t - y \sin \omega_0 t$$

where A is the unmodulated amplitude. Then, the envelope and phase are

Envelope: $\qquad r^2 = (x + A)^2 + y^2 = x'^2 + y^2$

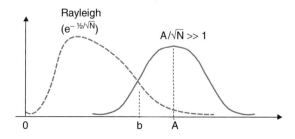

Figure A.1. The Rayleigh distribution.

Phase: $\theta = \tan^{-1}(y/(x+A)) = \tan^{-1} y/x'$

Then, the distribution of the envelope is

$$P(r) = (r/N) \exp(-r^2/2N) \exp(-A^2/2N) I_0(rA/N)$$

Where $I_0(z)$ is the unmodified Bessel function described as $I(z) = (1/2\pi) \text{Int}[e^{z \cos \theta} d\theta]$, where the integral is from 0 to 2π.

In binary form, a threshold is set at level b. Then, the probability for error is found by integrating from 0 to b:

$$P(r) = \exp(-A^2/2N) \text{Int}\{(r/N) \exp(-r^2/2N) I_0(rA/N)\} dr$$

and the Q-function is found by integrating from b to infinity:

$$Q(a, b) = \text{Int}\{x(\exp[a^2 + x^2)/2]) I_0(ax)\} dx$$

Then, the probability is a function of the average power signal-to-noise ratio γ and the normalized voltage level for detection b_0 (which depends on the receiver in use):

$$P(r) = 1 - Q(\sqrt{2\gamma}, b_0)$$

For commercial receivers the output signal-to-noise ratio may be approximated to

$$S_0/N_0 = \tfrac{1}{2}[\gamma^2/(1 + 2\gamma)]$$

APPENDIX B
Exercises and Simulation Examples Contained in the CD-ROM

The CD-ROM that accompanies this book contains ten characteristic examples that demonstrate how the quality and integrity of the WDM signal and performance of the optical link are affected by varying specific parameters through monitoring the power spectra, SNR, eye diagram, Q, and BER. One example among them demonstrates the effect of FEC on link reach (length) for an expected BER performance. The user is able to interactively simulate various conditions and verify how the optical power of the signal changes, or how dispersion affects the signal as the length of the fiber changes, and so on, as a result of changes in specific parameters. Visualizers allow the user to view the shape of the received pulse, the spectrum of the signal at the Demux, the eye diagram, BER, and more. The simulation tool used for the exercises is *OptSim*™ from RSoft Design Group, Inc. The simulation tool is intuitive and no prior knowledge of it is necessary. Documentation on the tool as well as detailed descriptions of the exercises are contained on the CD-ROM along with the software *OptSim*™. This demonstration version of *OptSim*™ was developed by RSoft Design Group, Inc. Further information on the most recent full versions of this and other RSoft products may be obtained by contacting RSoft Design Group.*

The following is a list of the exercises that have been designed to provide a realistic "look and feel" of fiber optic communication systems. We hope that these examples add value and enhance your understanding of the subject. Installation instructions are at the and of this appendix.

1. Demonstration of the effect of fiber length on received optical power. A four-channel WDM link at 10 Gbit/s per channel demonstrates how the fiber length affects received optical power, received eye diagram, optical power spectra prior to Demux, Q, and BER. There are three different topology files, each simulating a different fiber length. The example includes a discussion of how the various fiber effects impact the signal and system performance.

2. Demonstration of the effect of A_{eff} of a single channel link over single-mode

---

*RSoft Design Group, Inc., 200 Executive Blvd., Ossining, NY 10562. Voice: (914) 923-2164. FAX: (914) 923-2169. Email: info@rsoftdesign.com. Web: www.rsoftdesign.com.

fiber on received power, optical signal spectra, eye diagram, Q, and BER at the receiver. There are three different topology files, each simulating a different core-effective area.
3. Demonstration of the effect of dispersion on a single channel link at 10 Gbit/s over single-mode fiber. In the model, nonlinearities are neglected and turned off. There are three topologies, each simulating a different dispersion. All topologies use the same wavelength. The effect of dispersion on a signal for various link lengths of single-mode fiber in terms of signal spectra, eye diagram, Q, and BER at the receiver are shown.
4. Demonstration of the effect of dispersion on a 1 Gbit/s system versus a 10 Gbit/s system with a single channel link. There are four topology files, two with different dispersion coefficients and two with different bit rates.
5. Demonstration of the effect of dispersion compensation on received optical signal spectra, received eye diagram, Q, and BER.
6. Demonstration of the effect of polarization-mode dispersion (PMD) on received eye diagram, Q, and BER at different bit rates. There are two topology files for two different bit rates.
7. Demonstration of the effect of channel separation and FWM on received optical power, signal spectra at Demux, Q, and BER. A 10 Gbit/s four-channel WDM link is used to demonstrate the FWM effect. There are three topology files, each of which has a different combination of channel spacing and input power.
8. Demonstration of the effect of 10 Gbit/s modulation (NRZ versus RZ) on received optical power spectra, eye diagram, Q, and BER. There are two topology files, one for each modulation format.
9. Demonstration of the effect of EDFA amplification of a four-channel WDM link with different channel spacing on signal spectra before Demux, OSNR, received eye diagram, Q, BER, and EDFA model output plots. There are two topology files, each simulating a different channel spacing.
10. Demonstration of the impact of forward error correction (FEC) on system reach. There are two topology files, one for a system without FEC meeting an expected performance level (e.g., BER = 10E-9) at a length L1, and another with Reed–Solomon FEC meeting the same performance at another length L2.

Installation Notes for *OptSim*™ Demonstration Software

The demonstration software provided on the CD-ROM is included with this book. This software is for use on Windows 2000/XP computers only. To install this software, you must be logged into Windows as user name administration. Then run the setup.exe program included on the CD-ROM. You must agree to the License Agreement presented during the installation process to perform the installation. If you do not agree, you must cancel the installation and not install the software. The installation program will guide you through the installation steps.

Please read the readme file(s) on the CD-ROM and the installation directory after the software is installed for the latest detailed information about the demonstration software. There may be updates to the demonstration software available on RSoft's Web site. To check for these, go to RSoft Design Group's Web site at www.rsoftdesign.com/demo.

Index

1/f noise, 162
3-dB bandwidth, 93

Absolute efficiency, 111
Absolute spectral power responsivity, 92
Absorption, 37, 40
Allan deviation, 194
Amplified spontaneous light emission (ASE), 117
  backward, 117
  forward, 117
Amplifier noise, 164
Amplitude, 12
Amplitude modulated signals, 10
Amplitude modulating function, 15
Amplitude modulation, (AM), 13, 15
Amplitude shift keying (ASK), 12
Analog mobile communications system (AMPS), 7
Analog signal, 7
Angular dispersion, 110
Anisotropically optically transparent matter, 36
Anisotropy, 36
Asynchronous data networks, 183
Asynchronous networks, 1
Asynchronous transfer mode (ATM), 1
Attenuation, 9, 10, 58, 169, 170, 227
Attributes of matter, 36
Avalanche photodiode (APD), 93, 94
  deep-diffusion type, 94
  reach-through type, 95
  superionization type, 95

Background, 52
Background block error (BBE), 255
Bandwidth, 114
Bandwidth management, 136

BBE Ratio, 255
BER circuit, 254
  performance, 255
BER statistical measurements, 249
Binary BCH codes, 270
Binomial distribution, 224
Birefringence, 37, 50
Bit error contributors, 227
Bit error monitoring, 239
Bit error probability, 226
Bit error rate (BER), 13, 73, 214, 215, 228
  estimating, 252
  in-service, 215oOut-of-service, 215
Bit error ratio, 214
  penalty criterion, 206
Bit spreading, 227
Bit synchronization, 184
Bit-rate limits, 13
Bits, 1
Block codes, 266
Block error ratio, 216
Blue shift, 40
Bragg grating temperature, 112
Brewster's law, 47
Bright zones, 34
Brillouin frequency, 122
Brillouin scattering (BS), 54
Burst errored seconds (BES), 255
Bytes, 1, 3

Carrier, 16
Carrier-to-noise ratio, 217
C-band, 9
Channel capacity, 234
Channel width, 105
Chromatic dispersion, 64, 68
Chromatic dispersion coefficient, 66
Chromatic jitter, 167, 202

Circularly polarized wave, 46
Cladding, 55
Classical interference, 33
Cliff, 237
Clock accuracy, 181, 182, 187
Clock stability, 187
Code interleaving, 260
Coherent detection, 96
Coherent heterodyne detection, 22, 96
Computer communications, 1
Computer network, 1
Connecting fiber ends, 103
Constructive interference, 34
Continuous wave (CW), 11
Convolutional codes, 266
Copper-based legacy network, 1
Critical angle, 38
Cross talk, 75, 166, 227
Cross-gain modulation, 123
Cross-phase modulation (XPM), 123, 128, 228
Cut-off frequency, 93
Cut-off wavelength, 58
Cyclic codes, 269

Damping factor, 10
Dark current, 93
Dark zones, 34
Data communications network, 1
Data network, 1
Data patterning, 239
Data-dependent jitter (DDJ), 200
dBm, 75
Decibels (dB), 75
Demodulation of optical signal, 97
Dense wavelength division multiplexing (DWDM), 29
Dependence of gratings on temperature, 112
Destructive interference, 34
Detection, 86, 87
Detection techniques, 96
Dielectric constant, 32
Differential group delay (DGD), 70
Diffraction, 41
　by reflection, 44
　by transmission, 43
　gratings, 109
Digital modulation, 9
Digital pulse-coded modulation (PCM), 7

Digital signal level 0 (DS0), 7
Digital transmission, 7
　versus analog transmission, 7
Dimensioning, 3
Direct detection, 14
Direct modulation, 14
Dispersion, 62, 70
Dispersion compensation, 68
Dispersion limits, 13
Dispersion slope, 68
　mismatch, 68
Dispersion-shifted fiber (DSF), 67, 68
Dopants, 54
Double Rayleigh scattering (DRS), 121
DS1, 7
DS3, 7
Duo-binary encoding, 17, 18
Duo-binary level, 19
Duo-binary modulation, 17
Duo-binary signal, 18
DWDM network restoration, 134
DWDM optical network, 128

Echo, 152
EDFAs, 116
Effective area, 57
Effective generated power (EGP), 4
Einstein's energy relationship, 28
Electric dipole moment per unit volume, 44
Electric polarization, 32
Electric susceptibility of medium, 32
Electroabsorption, 20
Electro-absorption (EA), 21
Electroabsorption multiquantum well (EA-MQW), 20, 21
Electromagnetic fields, 33
Electromagnetic wavelet, 28
Electron, 27
Electrooptic crystals, 16
Electrorefraction, 20
Electrorefraction modulation (ERM), 16
Electrorefraction modulators (ER), 21
Elliptically polarized wave, 46
Error detection and correction codes, 259
Error detection and correction strategies, 260
Errored block (EB), 216, 255
Errored second (ES), 216, 255
Errored second ratio (ESR), 216

ES satio (ESR), 256
Extinction ratio, 51
Extraordinary index, 50
Extraordinary ray, 50
Eye diagram, 240
Eye-pattern mask, 243

Factors affecting matter and light, 128
Fano factor, 161
Faraday effect, 107
FEC code, 270
Fiber attenuation, 58
Fiber birefringence, 61
Fiber Bragg grating (FBG), 112
Fiber loss, 58
Fiber modes, 55
Fiber-span limit, 13
Figure of merit (FOM), 95
Filter, 10
   bandwidth, 10, 11
   response, 10
   slope, 106
Flicker noise, 162
Forward and backward defect indication (OCh-FDI/BDI), 135
Forward-biased noise, 93
Four-fiber bi-directional line switched ring (4f-BLSR), 132
Four-photon mixing, 74
Four-wave mixing (FWM), 73, 74, 123, 128, 166, 203, 228
   temporal, 126
Frame relay (FR), 1
Frame synchronization, 184
Fraunhofer diffraction, 41, 42
Free-running accuracy, 182
Frequency, 12
   bandwidth, 93
   distortion, 151
   modulation index (FMI), 17
   shift keying (FSK), 12, 17
   window, 10
Fresnel equation, 38
Fresnel reflection, 38
Fringes, 33
FSK demodulation, 24

Gain, 10, 114
   efficiency, 114
   saturation, 114
Gaussian distribution, 225
Generalized MPLS (GMPLS), 130
Generic framing procedure (GFP), 130
Glass, 29
Glass fiber, 29, 54
Gratings, 109
Group delay rate change, 66
Group velocity, 39, 40
Group velocity dispersion (GVD), 66

Hamming codes, 263
Helmholz equation, 41
Heterogeneously optically transparent matter, 36
Histograms, 250
Holdover stability, 182
Homodyne detection, 22, 96
Homogeneously optically transparent matter, 36
Huygens's principle, 41
Huygens–Fresnel principle, 33

Index of refraction, 37, 38
Indium phosphide (InP), 20
Infrared band, 9
Insertion loss (IL), 101, 108
Intensity modulation, 15
Intensity modulation with direct detection (IM/DD), 12, 96
Internet Protocol, 1
Intersymbol interference (ISI), 17, 200
Isolation, 102, 108
Isotropic matter, 171
Isotropically optically transparent matter, 36
ITU-T G.692, 9
ITU-T Nominal Center Frequencies, 9

Jitter, 170, 181, 196, 199
   filtering templates, 204
   generation, 203
   maximum tolerable, 205
   self-induced, 202
   signal affected by, 201
   sources of, 201
   tolerance, 203
   transfer, 203
Jones vector, 45, 48

Kerr effect, 71, 72

Laser semiconductor device, 11
Lasers, 83
L-band, 9
Left-circularly polarized wave, 46
Legacy rates, 8
Light, 33
  attributes, 35
  from a source at infinity, 41
Light–matter interaction, 27
Limits of optical power in fiber, 60
Line spacing, 105
Line width, 105
Linear contributing effects, 13
Linear dispersion, 110
Linear response, 10
Linearly polarized wave, 45, 46
Link budget, 4
Link capacity adjustment scheme (LCAS), 130, 131
Link loss (LL), 4
Lithium niobate (LiNbO3), 20
Logic one, 1, 9
Logic zero, 1, 9
Loss, 150
Loss of clock (LOC), 189
Loss of signal (LOS), 189

Mach–Zehnder (M-Z) configuration, 14, 20
Magnetic field, 27
Material dispersion, 53, 54, 65
Matter, 35
Maximum optical path penalty (MOPP), 86
Maximum Time Interval Error (MTIE), 192, 194
Maximum tolerable jitter (MJT), 205
Maxwell's electromagnetic plane wave equations, 30
Mechanical pressure, 41
Metropolitan ring networks (Metro), 131, 132
Microcracks, 40
Minor principal transmittance, 51
Modal dispersion, 62, 63
Mode of polarization, 45
Modulation, 7, 11, 13, 14
Modulation index, 15

Modulation instability, 73, 169
Modulation instability (MI), 128, 228
Modulator types, 20
Momentum, 33
Monochromatic light, 30
Multimode fiber, 55, 56
multimode-fiber graded index (MMF GRIN), 56
Multiple optical channels, 123
Multiplex, 7
Multiquantum well (MQW), 14

Natural bit rate, 3
Negative helicity, 46
Network hierarchy, 189
Network synchronization, 183
Next-generation DWDM network, 129
Next-generation SONET/SDH, 130
Noise, 114, 147
  accumulation, 228
  extrinsic, 148
  intrinsic, 148
  sources, 147
Noise figure (NF), 164
Noise sources affecting the optical signal, 147
Noise-equivalent power, 93
Nonlinear contributing effect, 13
Nonlinear phenomena, 124
Nonoptically transparent matter, 36
Nonreturn to zero (NRZ), 13, 20
Normal distribution, 225
NRZ Versus RZ, 13
Number of fibers, 4

OFA amplification, 121
On–off keying (OOK), 12, 13, 16, 51
OOK modulation, 20
OOK NRZ, 23
Opacity, 105
Opaque matter, 36
Operating parameters, 4
Optical amplifiers (OA), 113, 114
Optical attenuators, 36
Optical channel isolation, 102
optical channel separation, 102
Optical circulators, 108
Optical coherent methods, 22
Optical communication systems, 21

Optical communications, 9, 11, 13, 15, 16, 19, 29, 86
Optical decoding, 21
Optical demultiplexers, 104
  active, 104
  passive, 104
Optical density, 105
Optical fiber amplifiers (OFAs), 114, 115
  line amplifiers, 115
  power amplifiers, 115
  preamplifiers, 115
Optical fiber network, 7
Optical filters, 36, 104
  fixed, 105
  tunable, 105
Optical frequency, 36
Optical intensity, 36
Optical isolators, 108
Optical multiplexers, 104
Optical noise, 227
Optical nonlinearities, 227
Optical power, 13
Optical power attenuators (OPAs), 107
Optical propagation, 27
Optical receivers, 81, 84
Optical semiconductor modulators, 20
Optical signal-to-noise degradation, 75
Optical signal-to-noise ratio (OSNR), 148, 217, 219, 233
Optical transmission medium, 54
Optical transmitters, 81
Optical transport network (OTN), 133
Optical-fiber-based network, 1
Optical-frequency shifting, 123
Optically transparent matter, 35, 36
Optical-path penalty (OPP), 86
Optical-performance monitoring, 137
Optimum dispersion, 9
Ordinary index, 50
Ordinary ray, 50
Oscillator noise, 149
Output response, 10
Output saturation power, 114
Oversampling, 99
Overshoot, 10, 11

Packets, 1, 3
Paradoxes, 52, 53
Parameters of the photonic signal, 11
Parity, 260
Particle nature of Light, 33
Pass-through grating, 109
Path engineering, 4
Path protected mesh network (PPMN), 133
Payload synchronization, 184
Per-channel automatic protection switching (OCh-APS), 135
Per-channel signal quality monitoring and backward quality indication (OCh-SQM/BQI), 136
Per-channel tandem connection maintenance (OCh-TCM), 136
Phase, 11
  change, 40
  difference, 17
  distortion, 150
  shift, 37, 52
  velocity, 39
Phase-lock loop (PLL), 182
Phase-shift keying (PSK), 12, 16, 19
Phase-shift keying signal, 16
Phenomenon of Fresnel, 41
Photodetector figure of merit, 95
Photodetector responsivity, 173
Photodetectors, 91
Photon, 27, 28, 34, 35
Photonic beam, 11
Photorefraction, 107
Photorefractive medium, 107
Photosensitivity, 92
Physical layer devices, 3
PIN photodiodes, 93
p–n diode semiconductor material, 91
Poincaré sphere, 49, 70
Poisonian power noise, 162
Poisson distribution, 225
Polarization, 12, 37, 44, 228
  by reflection, 47
  distortion, 152
  diversity, 50
  mode dispersion coefficient, 69
  mode shift, 107
  sensitivity, 114
  spreading, 50
  vector, 44
Polarization-dependent loss (PDL), 50, 51, 71, 228

Polarization dispersion mode (PMD), 20, 69, 167, 228
Polarization hole burning (PHB), 228
Polarization mode coupling (PMC), 70
Polarization-mode dispersion compensation, 70
Polarizing filter, 36
Positive helicity, 46
Positive intrinsic negative (PIN) photodiode, 93
Power attenuation, 58
Power loss, 58
Principal transmittance, 51
Principles of decoding, 21
Probability and statistics, 220
Probability density function (PDF), 223
Probability theory of bit error rate, 213
Propagation of light, 37
PSK, 20
 demodulation, 24
Public switched telecommunications network (PSTN), 1
Pull-in/hold-in, 183
Pulse amplitude modulation (PAM), 14
Pulse broadening, 73
Pulse shape, 98
Pulse timing, 70
Pulse-width eistortion jitter (PWDJ), 200

Q-factor, 237
Quality factor, 237
Quantization error, 217
Quantization noise, 217
 improvement, 235
Quantum efficiency, 93, 94
Quantum interference, 34
Quantum interferometry, 35
Quantum limit, 84

Raman amplification, 117
Raman amplifiers, 114, 117, 121
Raman figure of merit (FOM), 121
Raman noise, 119
Raman supercontinuum, 119
Receiver penalty, 13
Receiver signal level (RSL), 4
Red shift, 40
Reed–Solomon Codes, 270
Reflection, 37

Reflection grating, 109
Reflectivity, 37
Refraction, 37
Refractive index, 36, 38, 40
Responsivity, 94
Return to zero (RZ), 13, 20
Right-circularly polarized wave, 46
Rise time, 11
Rise/fall, 10
Rotators, 107
Round-trip delay, 152

Sampling, 98, 239
Sampling methods, 99
Scattering, 37, 40
Self-induced optical noise, 202
Self-modulation, 73, 169
Self-phase modulation (SPM), 71, 72, 128, 168, 228
Semiconductor laser, 11
Semiconductor optical amplifiers (SOAs), 114
Semitransparent matter, 36
Sensitivity, 93
Service-level agreements (SLAs), 129
SES Ratio, 256
Severely errored second (SES), 216, 255
Severely errored second ratio (SESR), 216
Shannon's limit, 219
Shot noise, 159, 161, 162
 electronic, 159
Sideband frequency, 15
Sidetones, 227
Signal detection, 13
Signal label (OCh-SL), 135
Signal quality, 195
Signal-power depletion, 75
Signal-to-noise ratio (SNR), 13, 148, 206
Silica core, 54
Singing, 153
Single mode fiber, 56
Sinusoidal jitter (SJ), 201
Snell's law, 37, 38
SONET synchronization, 185
SONET/SDH rates, 8
Spectral bandwidth, 17
Spectral broadening, 40, 168
Spectral matching, 216
Spectral monitoring, 239

Spectral resolution, 111
Spectral response, 92
Spectral width, 105
Spectrum, 9
Specular reflection, 109
Speed of light, 33
   in free space, 38
   in medium, 38
Square input pulses, 10
Standard single mode fiber (SSMF), 56
State-of-polarization shift keying (SoPSK), 12, 19
Statistical Soise, 149
   in communications, 277
Statistical sampling for BER, 251
Steepness, 106
Stimulated Brillouin scattering (SBS), 122, 127, 227
Stimulated Raman scattering (SRS), 54, 127, 227
Stokes noise, 167, 202
Stokes nector, 47, 48
Synchronization impairments, 188
Synchronous hierarchy, 7
Synchronous networks, 1

Telegraph, 1
Telephony, 1
Temperature variation, 41, 228
Temporal FWM, 126
Temporal parameters, 169
Terminal capacitance, 93
Termination, 239
Thermal (or Johnson) noise, 149
Thermal noise, 153
   at the Receiver, 158
Time deviation (TDEV), 192
Timing, 181
   error, 190

protection, 187
reference source, 188
response, 93
Topologies, 131
Traffic probability, 195
Trail trace (OCh-ID), 135
Transfer function, 10
Transmission engineering, 3
Transmittance, 105
Transparency, 37
Transportable amount of information per channel, 13
Travel distance, 34
Two-fiber bi-directional line switched ring (2f-BLSR), 131

Uncorrelated bounded jitter (UBJ), 201
Undershoot, 10, 11

Verdet constant, 107

Wander, 181, 196, 207
   signal affected by, 207
   sources of, 207
Wave nature of light, 30
Waveguide dispersion, 65
Wavelength collision, 136
Wavelength conversion, 122
Wavelength converters, 122
wavelength dispersion, 65
Wavelength management, 136
Wavelengths per fiber, 4
Wavelet, 28
WDM, 14

Zero-dispersion wavelength, 40, 67
Zero-order diffraction, 109

CUSTOMER NOTE: IF THIS BOOK IS ACCOMPANIED BY SOFTWARE, PLEASE READ THE FOLLOWING BEFORE OPENING THE PACKAGE.

This software contains files to help you utilize the models described in the accompanying book. By opening the package, you are agreeing to be bound by the following agreement:

This software product is protected by copyright and all rights are reserved by the author and John Wiley & Sons, Inc. You are licensed to use this software on a single computer. Copying the software to another medium or format for use on a single computer does not violate the U.S. Copyright Law. Copying the software for any other purpose is a violation of the U.S. Copyright Law.

This software product is sold as is without warranty of any kind, either express or implied, including but not limited to the implied warranty of merchantability and fitness for a particular purpose. Neither Wiley nor its dealers or distributors assumes any liability of any alleged or actual damages arising from the use of or the inability to use this software. (Some states do not allow the exclusion of implied warranties, so the exclusion may not apply to you.)

WILEY